Timber and Iron Reinforcement in Early Buildings

By R. P. Wilcox

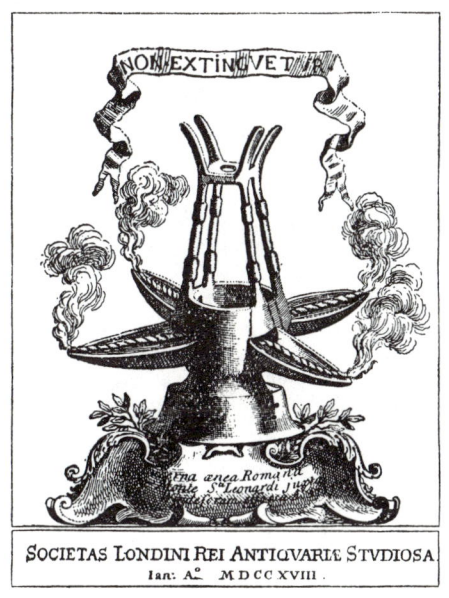

Occasional Paper (New Series) II

THE SOCIETY OF ANTIQUARIES
OF LONDON

Burlington House, Piccadilly, London W1V 0HS
1981

TYPESET BY BISHOPSGATE PRESS
PRINTED BY WHITSTABLE LITHO

Timber and Iron Reinforcement in Early Buildings

The use of reinforcement in building has a very long history. Man, in his many attempts at building construction, has often seen his structures suffer from agents of destruction, some natural, some human, and he has made various attempts to augment his constructional materials with materials of another kind in order to give them resistance to stresses and strains. These destructive factors have three main causes: the action of violent earth movements (earthquakes), the stresses inherent in the settling of a building immediately after completion, and the attempts of enemies to destroy the structure.

Three types of reinforcement will be discussed in this article. The first, intra-mural timber reinforcement, was embedded in the walls of structures, usually in the cores of masonry walls either as beds into which to joint floor joists or as bonding timbers to provide longitudinal stability, in direct contrast to buttresses which gave stability in the opposite direction. The second, foundation reinforcement, utilised timbers in the foundations of buildings in order to provide a solid raft on which the building could be erected. The third, obvious to the casual eye, was the use of timber and iron reinforcement in the interior of buildings, for instance joining opposing piers across the naves of churches, or tying together the haunches of arches in a nave arcade.

In some cases, two or more of these techniques were used in a single building or structure, but, more commonly, they were employed individually and, in the interests of clear exposition, they will be treated here as separate strands in the history of architecture. But, as will appear later, two of the techniques were originally responses to the same destructive threat.

Intra-mural timbers

In the Near East and Europe where civilization flourished during the prehistoric period, early examples of the use of this type of reinforcement can be drawn from buildings constructed of masonry and mud-brick but, outside these areas, in the

1

barbarian parts of Europe, buildings were not of stone or mud-brick in earlier times, but of timber, and the only use of reinforcement was in the ramparts of fortifications of the hill-fort type. The earliest belong to the ninth century B.C. and are found in central Europe. In Switzerland the Wittnauer Horn (Bersu, 1946), a well excavated example, had a rampart constructed in different ways at two distinct periods. During the first period, a sloping-faced rampart was laced with horizontal timbers, and in the second period (c. 700B.C.) a vertical-faced rampart was braced with vertical timbers. Not far away, at the same time, a hill-fort was being constructed on the Montlingerberg near Oberrit (Frei, 1954–55) with a rampart composed of timber-framed boxes on stone bases packed with clay. Later, in England, timber-laced ramparts appear in Sussex in the fifth century B.C. at Hollingbury (Curwen) and Ranscombe (Burstow and Holleyman), and a hundred years or so afterwards in Scotland in stone ramparts which later became vitrified (Cotton, 26–105). Timber-lacing has been used as part of the basis of a scheme of development for hill-forts in Britain postulated by Professor Cunliffe (Cunliffe, 1974).

From the middle of the second century B.C., we can find examples of Celtic *oppida* on the Continent. These were surrounded by ramparts that were constructed of earth and stone with timber-lacing. Two building techniques have been recognized (Piggott). The earlier is at Preist, Trier in Germany, where the double drystone walling faces are tied together with transverse beams and filled in with earth. Upright beams revet the front wall and support with their upper ends a timber breastwork mounted on the front edge of the rampart. In the first century B.C., as a defence against Roman siege tactics, the technique changed and nails were used to fasten the timbers together at their intersections, with additional timbers running longitudinally through the width of the rampart. This was the 'murus gallicus' of Caesar[1], but his description is not particularly precise and may have been written from secondhand information. Examples of timber-laced walls that have been excavated, like those at Manching (Krämer), do not tally completely with the purely longitudinal nature of the beams in Caesar's description, and appear to have contained transverse members as well.

In Britain, in the later pre-Roman Iron Age, there are no close analogies to the timber-laced ramparts on the Continent. The later periods at Mount Caburn, Sussex (Wilson, 1938 and 1939), and at Maiden Castle (Wheeler, 1943) have produced some evidence of timber-lacing, but hill-forts at this time north of the English Channel seem to have relied mainly on dump ramparts surmounted by a palisade.

The use of timber-lacing in ramparts does not seem to have been originally introduced in response to a new form of attack; one cannot imagine that there would have been any form of attack in the late Bronze Age that could not have been as well deterred by a dump rampart. But what intra-mural timbers could do was to provide a more imposing rampart and they involved the builders in a more architectural design than had hitherto been employed. The rampart could be higher than many dump ramparts; it could have vertical faces kept upright by vertical posts and stone walls; the faces could be tied into the body of the bank or to each other by horizontal timbers; and the rampart itself could be given a square section with better provision for wall-walks, palisades and gateways. No doubt timber-lacing was something of a status symbol, and certainly it was expensive

(Piggott), but it shows that a tradition of using timber encased in other materials was established in barbarian times before the coming of the Roman armies. The Romans took over the technique and used it occasionally. There are Roman examples at Caerleon in the Flavian rampart (Nash Williams), Coelbren (Morgan) and at Montgomery (Forden Gaer) (Price and Price) in Wales, but it is not very common and the usual Roman method of defence was either by a dump rampart or, later, by masonry walls.

To investigate the earliest use of timber in masonry or brick, one must turn to the Near East and the architecture of Mesopotamia and Egypt in the fifth millennium B.C. In the area of Mesopotamian influence, timber was embedded in brick construction as a reinforcement (Badaway), while in Egypt local woods such as acacia and date-palm were used in the same way (Badaway). In Asia Minor and northern Syria

> 'Forests must have been more extensive in antiquity, since timber from coniferous trees was freely used as reinforcement against frequent earthquakes. The wooden beams were set in the stone or sun-dried brick structures which had stone foundations and ceilings as well as corbel vaults of stone or true vaults of brick'. (Badaway, 120)

At Alishar Hüyük (Osten) beams were let in horizontally along both faces of walls, the beams varying from 0.25 to 0.3 m. (10 to 12 in.) thick. At Troy II, the four-metre (13-ft.) thick walls of sun-dried brick were reinforced with beams of 0.31 m. (12 in.) scantling running both longitudinally and transversely so that the ends of the transverse beams were flush with the faces of the wall. This was during the phase that came to an end in c. 2300 B.C. (Piggott). At Knossos about 1550 B.C. the Hall of the Double Axes was built of brick and horizontal timbers. In the back wall appears a course of cylindrical beams laid transversely so that the ends, flush with the interior face of the wall, must have appeared on the surface as a row of wooden discs (Bell).

Other examples show the use of a combination of longitudinal and transverse beams. The city walls of Early Bronze Age Jericho contained the charred remains of such a system (Kenyon). At Alalakh (Woolley, 1955) there were three tiers of both longitudinal and transverse beams in the original structure. Vertical beams were employed as well at other places. An example of c. 1500 B.C. was excavated at Beycesultan in Anatolia (Lloyd and Mellaart, 1955); vertical, longitudinal and transverse beams had been used in the walls. In Crete, the houses of Vasiliki, built of sun-dried brick, had their interior wall faces divided into rectangular compartments by timber beams. Wace reports that the Mycenaean structural system used in houses and palaces and in both brick and ashlar construction involved horizontal and vertical timbering. This is confirmed by Mylonas.

The technique of intra-mural reinforcement, in varying forms, seems to have been common to the whole of the prehistoric eastern Mediterranean: Lloyd and Mellaart vouch for its existence in the Aegean, Woolley (1954) describes its use in Iraq at Ur, and Koldeway at Babylon. The examples range in date from Neolithic times at Jericho down to Iron Age Zendjirli (Conteneau).

In all these examples, the walls, whether of brick or other materials, are of massive thickness and would need, one might think, little in the way of reinforcement. But at moments of earthquake, when the ground becomes unstable, it is necessary to have a longitudinal element present in the structure that can accommodate itself to a certain extent to the strain but, more important, hold the building together and prevent the unavoidable cracks from widening so much that the walls simply fall apart. Timber is by far the most useful material for this kind of reinforcement for it has an element of elasticity, but during later times in the Middle East, when timber was less abundant, masonry was used instead. In fortifications in the area there are examples of the use of stone pillars taken from other buildings. The Bāb An-Nasr gate in Cairo, built in 1087 (Cresswell), has columns set transversely in the walls so that their ends appear flush on the outer faces. This feature appears around all the walls of the gate and continues around the city walls as well. The same technique was employed at Alexandria where it was noticed by Gratien le Père who accompanied Napoleon's expedition in 1798. At Caesarea, in Palestine, the Crusaders built in 1091 a castle in which the masonry was reinforced with granite columns. Of this building an Arab writer said

'The Franks had transported columns of granite to this place, which they had placed transversely in the walls, so that they had not to fear sapping and could not fall should they be undermined' (Maqrīzī, 10–12)

The castle at Sheizar was rebuilt by Nur addin of Aleppo about 1160 after it had been destroyed by an earthquake. He attempted to protect it from future similar damage by reinforcing the walls with columns taken from the ancient settlement (Müller-Weiner). The Frankish fortress of Jebail north of Beirut had similar reinforcement in its curtain wall. It was built between 1197 and 1198 (Müller-Weiner).

It is clear that, by this later period, another function, that of reinforcement against siege operations, was being served by the intra-mural strengthening. Although such reinforcement was used in military structures in the West, it was employed more sparingly and in most cases could not have provided much protection against mining operations.

The use of timber in masonry was practised at Byzantium. Philo of Byzantium, in his work on fortifications written about 120 B.C., describes a use of intra-mural timbers that is similar to later examples in Europe. He advocates using baulks of oak buried in the body of both curtain walls and towers; the timbers being placed end to end and forming horizontal chains at vertical intervals of 2 m. (6 ft. 6 in.). Philo comments that the presence of these timbers greatly facilitates the repair of any part of the wall which may be damaged. A little later another architect mentions the use of intra-mural timbers in his manual of architecture and building construction. This was Marcus Vitruvius Pollio who was in Roman government service during the reign of the Emperor Augustus. He describes the technique employed in massive urban building of the time[2]. The walls of Strasbourg in Alsace are the nearest extant analogy to Vitruvius's description, but the two tiers of timbers are restricted to areas of the wall adjoining the towers and were used to tie the towers firmly into the town wall. In most cases the timbers themselves have long since rotted away leaving behind them the chases or slots they occupied and it is these hollows in the masonry that provide the bulk of the evidence for the use of timbers. Such evidence at Richborough Roman fort demonstrates that the towers there were tied into the fort walls in the same way as those of Strasbourg. Fig. 1

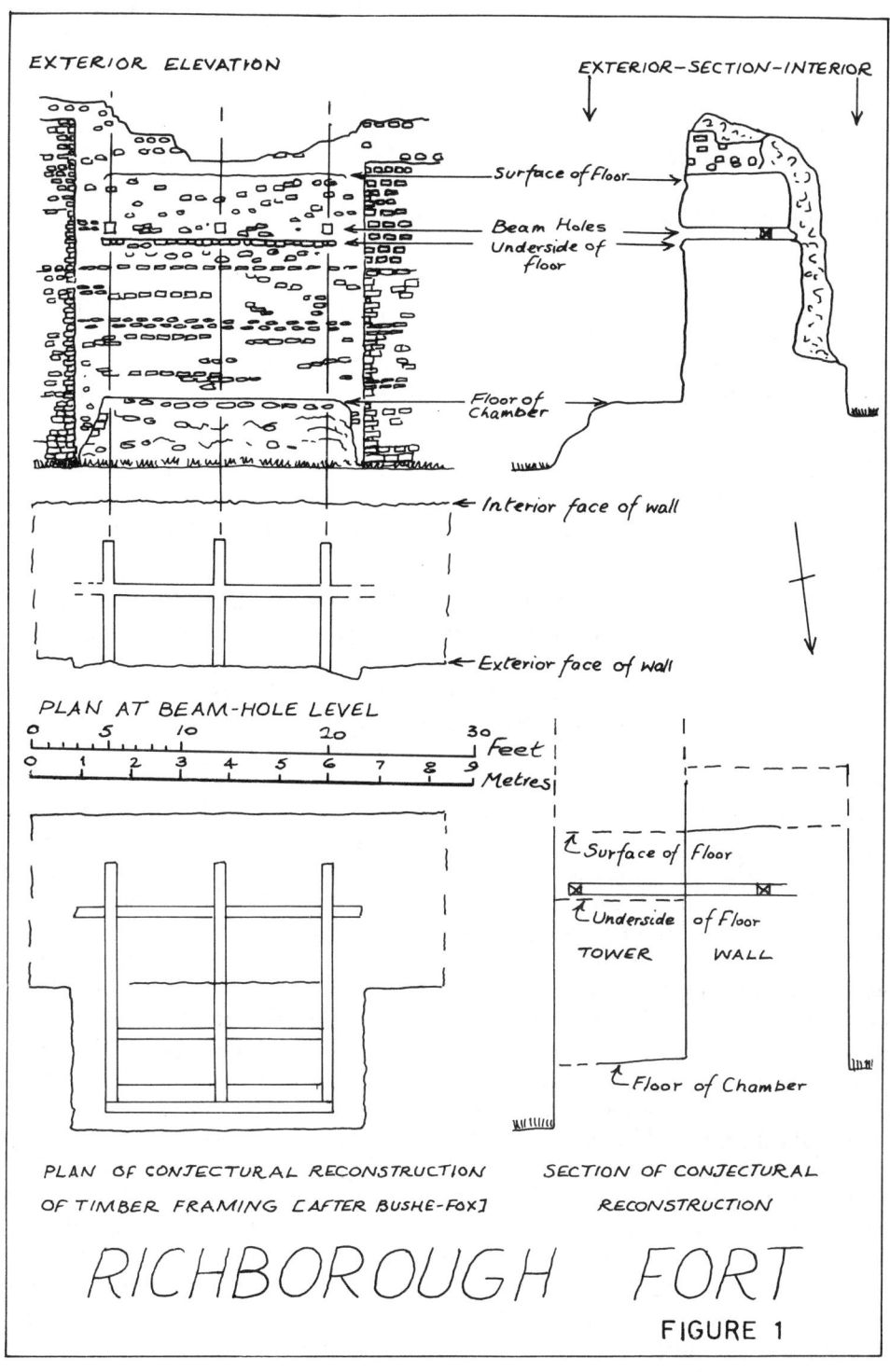

EXTERIOR ELEVATION

EXTERIOR—SECTION—INTERIOR

Surface of Floor

Beam Holes
Underside of floor

Floor of Chamber

← Interior face of wall

← Exterior face of wall

PLAN AT BEAM-HOLE LEVEL

Surface of floor

Underside of Floor

TOWER   WALL

Floor of Chamber

PLAN OF CONJECTURAL RECONSTRUCTION
OF TIMBER FRAMING [AFTER BUSHE-FOX]

SECTION OF CONJECTURAL
RECONSTRUCTION

RICHBOROUGH FORT

FIGURE 1

(after Bushe-Fox) also demonstrates that the beams served as reinforcement for the floors of the towers which may well have been used to carry light artillery.

In Britain, after this Roman example, there is a lacuna before the medieval uses referred to later on in this account, but on the Continent the construction of masonry buildings continued and with it the use of timber reinforcement. From the evidence of the surviving standing structures it is probably correct to say that the tradition continued in buildings created under the influence of Byzantine architecture. At Salonica, the walls of the city were reinforced with longitudinal baulks of timber placed at vertical intervals in the masonry (Choisy, 1883) which dates from the fourth to the fifth centuries A.D. In the ninth century the evidence comes from ecclesiastical buildings in widely-spread localities. In Anatolia, the church at Ala Klisse (Ramsey and Bell), built probably in the ninth century, has a corner overlooking a steep slope. This corner is reinforced with intra-mural timbers. Ohrid, in present-day Yugoslavia, contains the church of Sv. Sophia. Soon after the Second World War, the building was restored and the UNESCO report (Forlati) contains the following statement:

> 'There are many cavities in the walls where wooden ties or binders were formerly embedded in the masonry, but these have now rotted away and are worse than useless, constituting, as they do, dangerously weak spots in the structure' (Forlati, 9)

The nave, the part of the structure concerned, has been dated by a Yugoslavian authority (Stricĕvić) to the years between 852 and 859. The church of S. Ambrogio, Milan, was built in the early ninth century, but was reconstructed in the early twelfth century (Conant). However, one of the towers, the Campanile dei Monaci, was left untouched and still stands today. Its masonry is reinforced with intra-mural timbers.

Our evidence from the Roman period comes only from instances in masonry defences. Whether the technique was used in other forms of building is not yet clear. But, during the post-Roman period up to the beginning of the tenth century, in buildings which were influenced by Byzantine architecture, the examples are both defensive and ecclesiastical. In ecclesiastical architecture there is better continuity than in defensive structures: on the Continent four towers were constructed during the eleventh century, reinforced with intra-mural timbers like the tower at S. Ambrogio. S. Marco, Venice, stands fronting its famous tower in the Piazza di San Marco. The campanile was built a little later than the church, during the latter part of the eleventh century. It was reinforced with timber throughout (Choisy, 1883). In Como, the church of S. Carpoforo was built in the eleventh century (Porter, 1915–1917) and its tower was strengthened with intra-mural timbers and completed in 1070. Twenty-five years later, another church in Como, S. Abondio, was completed (Conant). Its northern campanile is modern but the other is of the same build as the rest of the church and is reinforced with timber beams. Also constructed in the eleventh century, further north in Burgundy, is the church at Baugy which has a tower built of rubble with bonding timbers (Oursel).

St. Ethelbert's Gate (fig. 2) leads into the cathedral close at Norwich. It is a Norman structure, remodelled in 1316 (Whittingham). Intra-mural timber chases were discovered during repairs to the gateway: running within the two walls parallel to the carriageway and about 3.66 m. (11 ft.) above present ground level

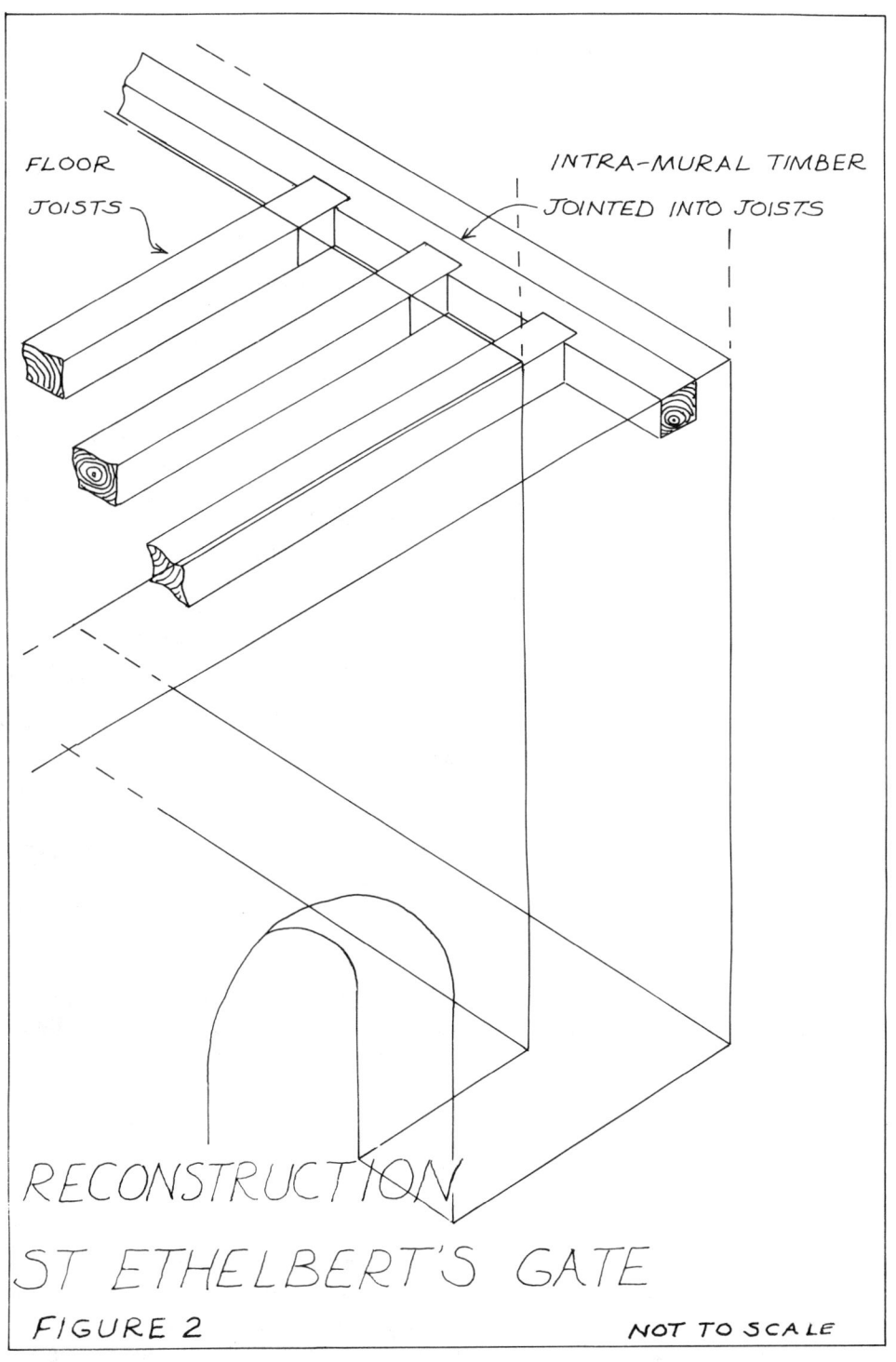

FLOOR JOISTS

INTRA-MURAL TIMBER JOINTED INTO JOISTS

RECONSTRUCTION
ST ETHELBERT'S GATE
FIGURE 2

NOT TO SCALE

were empty chases which originally held beams of approximately 20 by 22.5 cm. (8 by 9 in.) scantling. Into these beams were jointed floor joists of the chamber above the gateway. The joists were approximately of the same dimensions and it seems probable that they were half-jointed into the intra-mural timbers. When repair work was done, the empty wall chases were filled up with concrete and the cavities through which the joists entered the walls were outlined in the same material. The cathedral architect, Mr B. Feilden, has dated the original Norman gateway and therefore the intra-mural timbers to c. 1090.

The best account of the architectural history of Lewes Priory in Sussex is that by St John Hope and he has assigned the earliest building period there to between 1091 and 1110. Some buildings that remain in fragmentary condition today date from this period and one of these, the refectory, contains evidence of the use of intra-mural timbers. Only a small portion of this building is left standing beside the railway that was driven through the site but it contains a large intra-mural timber chase in the lower part of the wall (pl. 1). It is certain that this timber bonding was a good deal more extensive than it appears today, but it was not carried all the way round the building, for it does not appear at the other end of the fragment of wall in the photograph, neither does it appear in another fragment further to the east. At this height in the wall, it would have been interrupted by doorways, so that it could only have reinforced parts of the lower portion of the wall. Whether the ground in this area was unstable or not is impossible to determine today because most of the floor area and all the other walls disappeared during the railway excavations. But one might guess that stability was a problem in view of the fact that the scantling dimensions of the remaining timber chase is 30.5 by 34 cm. (12 by 13½ in.). Such a size indicates really massive reinforcement.

When in the 1140s the brotherhood had grown so large that it was necessary to enlarge the dorter, intra-mural timbers were used in the rebuilding. As the Priory had been built in rather a confined area on the crest of a rise, the new building had to be built down a slope whilst keeping the extension of the dorter on the same level as the original dormitory. As a result, the extension was raised on undercrofts built down the slope of the hill, each one being a little higher than its predecessor. The final undercroft housed the new rere-dorter, built at right-angles to the dorter block over a stream diverted for the purpose. This potentially unstable ground perhaps suggested the need for wall reinforcement. The first rere-dorter which was built over the stream flowing in its earlier bed higher up the slope had, as far as can be ascertained today, no intra-mural timber reinforcement, and it may be that the stream caused some instability in that building which prompted its insertion in the later building.

Fig. 3 shows the plan of the lower tier of intra-mural timbers in the new rere-dorter and it can be seen that they are not continuous. It is assumed that the upper tier ran all the way round the building apart from a doorway in the south wall and a doorway at the bridge entrance in the north wall at a height of approximately 4.5 m. (14 ft. 9 in.) above the present ground surface. The varying heights of the timbers in the lower tier cannot be accounted for by an examination of the walls as they stand today: it may be that the walls were built by different gangs of masons whose methods were not co-ordinated until they had reached the level of the upper tier of intra-mural timbers. The insertion of such timbers, as other evidence tends to show, was usually rule-of-thumb and it is only, as at

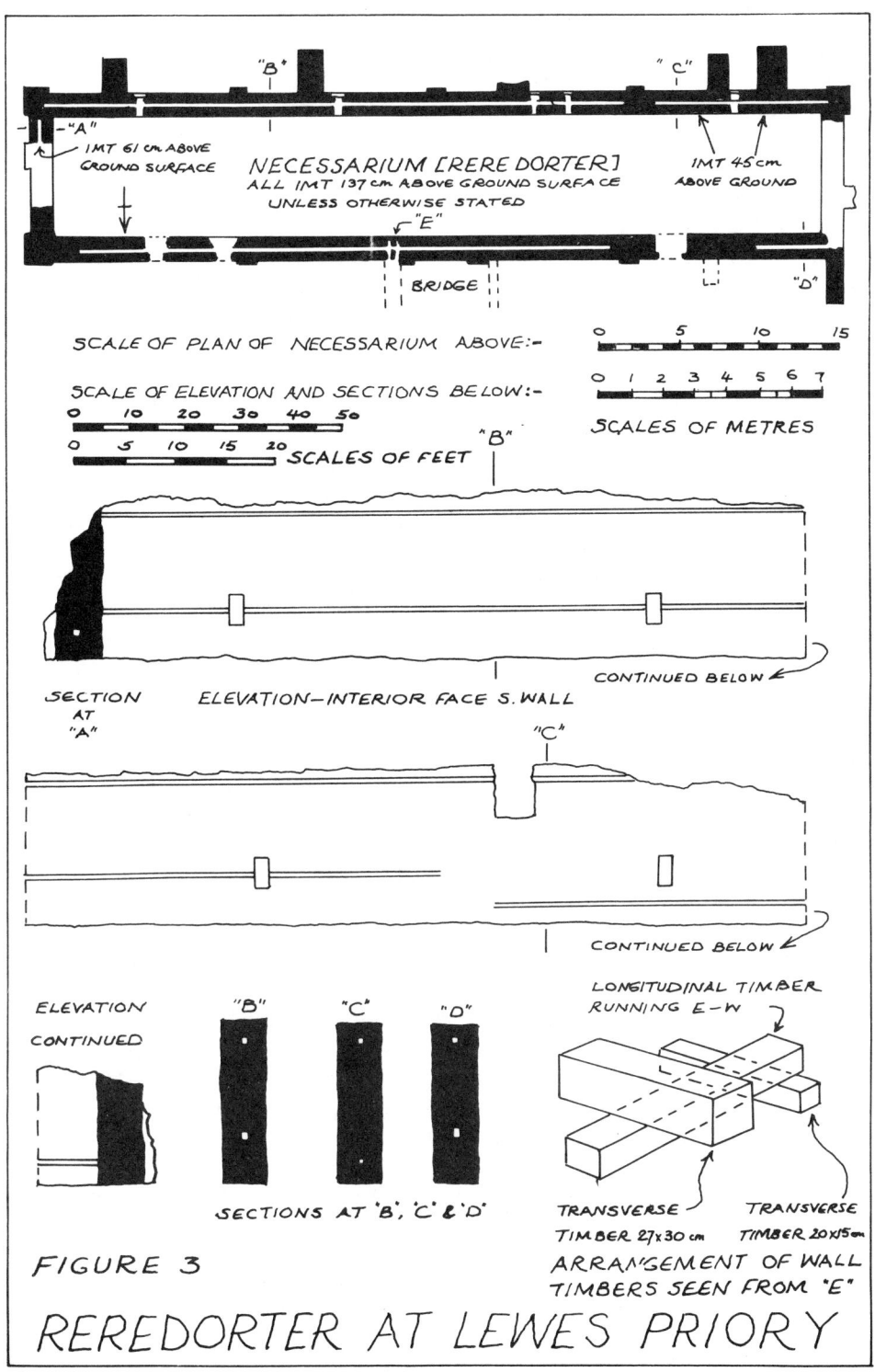

SCALE OF PLAN OF NECESSARIUM ABOVE:-

SCALE OF ELEVATION AND SECTIONS BELOW:-

SCALES OF METRES

FIGURE 3

REREDORTER AT LEWES PRIORY

Norwich's St. Ethelred's Gate and Bridgnorth Castle (described later in the account), where they are incorporated in a joist system, or in grander buildings, that they are organized on a regular plan.

In the rere-dorter at Lewes Priory, transverse beams are not associated with longitudinal ones except, in a few places, in the tier at the height of *c*. 1.42 m. (4 ft. 8 in.). Such a junction is best seen at point E (marked on the plan with a detail drawing below) where two transverse beams, which ran northwards through the wall of the bridge, met the longitudinal beam inside the north wall of the rere-dorter. The beams were either nailed or pegged together because, as can be seen in the detail drawing, the chases are quite separate and do not intersect as they would do if jointing had been employed. The upper tiers of intra-mural timbers which ran at an unvarying height through the walls are not linked to joist holes. The lower tier may have been so linked, but it is difficult to tell, for the joist holes in the rere-dorter are very irregular and by no means all original. After the dissolution of the priory the building was turned to other uses and different floors put in. These floors were not placed at the usual intervals, so that any inference drawn from the intervals between the joist holes would be unreliable.

The church of La Madeleine at Vézelay (Yonne) in Burgundy was completed about 1132 (Viollet-le-Duc). In the nave of the abbey church there was timber bonding (fig. 4a) above the archivolts giving on to the aisles and a second tier interrupted by the high windows at the level of the upper abaci of the columns at the springing of the high vaults. This second wooden bonding was attached to iron cramps which supported iron tie-rods spanning the nave at the base of the ribs. These tie-rods are a form of reinforcement which is discussed later in this article and this example best demonstrates the rare combination of these two forms of reinforcement. At first sight it would have appeared as a neat and satisfactory method of making a building into a perfectly rigid structure, but it obviously proved unsatisfactory, for the iron tie-rods lost their effectiveness with the decay of the intra-mural timbers and were later fixed with simple iron hooks into the masonry (fig. 4b).

The abbey of St. Denis near Paris had a tower built in the mid eleventh century which was demolished by Debret in 1846. In an account written before its demolition, there is a description of intra-mural timbers. Debret describes timber banding of large scantling fixed together with iron pegs and sunk within the four walls. The chases left after the decay of the wood were some 30 cm. (12 in.) in cross-section. Timbers that intersected in the middle of the tower were fixed to the centres of these intra-mural timbers at each stage and were used to bind together the four piers of the tower between the bays. These cross-ties were burnt out in the thirteenth century before the construction of a spire. This, like Vézelay, is an example of the combination of two forms of reinforcement with the cross-members of timber rather than iron in this case.

During the thirteenth century, there was an example of the use of intra-mural timbers at the Benedictine priory of Wilmington in Sussex where the hall had its south wall reinforced in this fashion. Two instances of approximately the same date occur in Northern Ireland. Both are in County Down and both are small rectangular churches. Maghera Old Church (Jope, 1966) had at least two intra-mural timbers; one through the base of the western gable at wall-plate level and another through the north wall. It is possible that the latter beam was jointed at

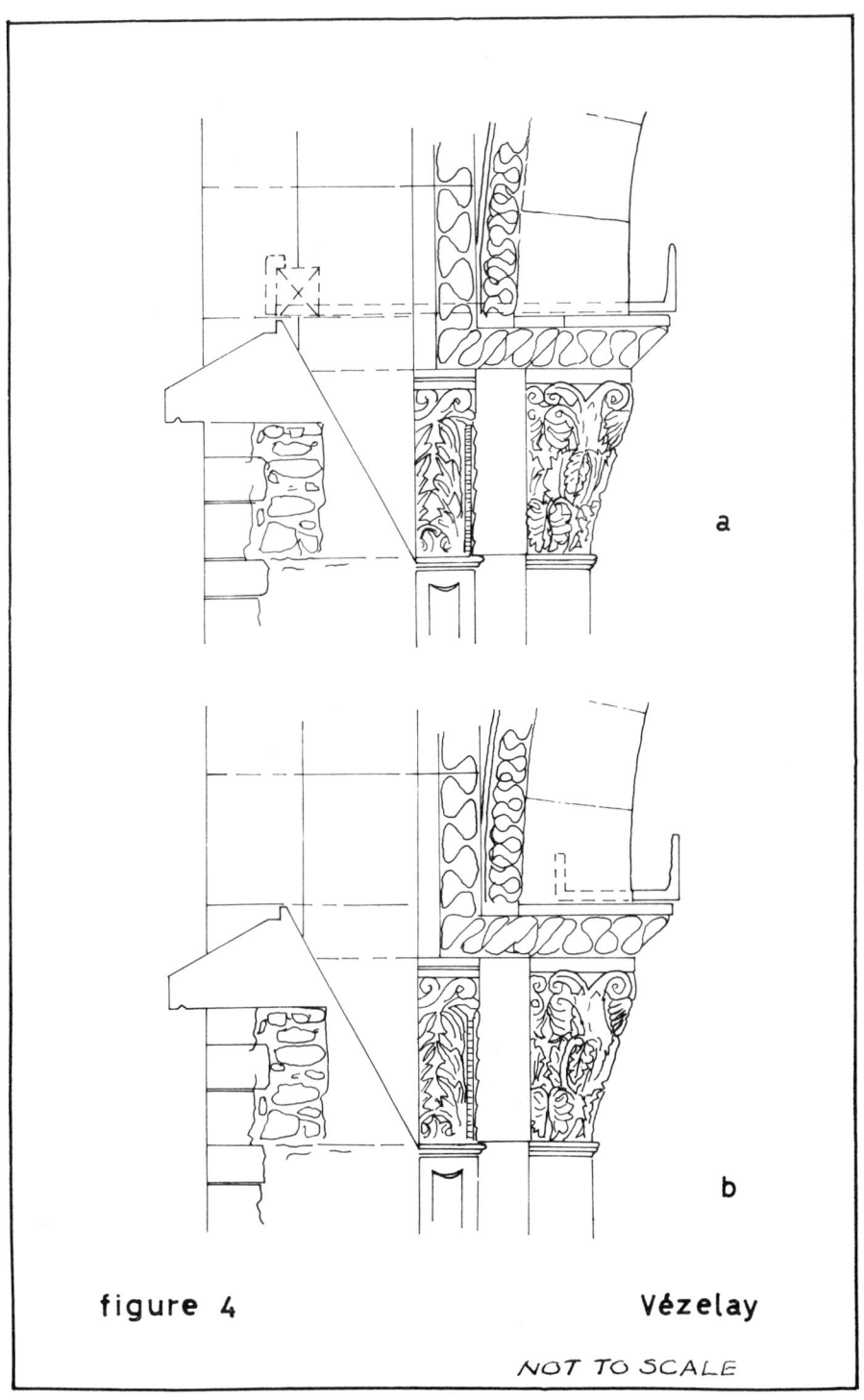

figure 4   Vézelay

NOT TO SCALE

right-angles into another running inside the east wall. The other church is at Derry South (Jope, 1966) where again an intra-mural timber was inserted at wall-plate level in the base of the western gable. It is connected with putlog holes that pierce the wall immediately beneath the timber chase.

Two monastic examples complete the list of ecclesiastical examples. Castle Acre Priory in Norfolk, a Cluniac monastery, has a range of ancillary buildings lying to the west of the main buildings. One of these still stands to a considerable height. At the east end, the north corner was reinforced with intra-mural timbers. At this point the building is close to the edge of the main drain of the monastery and is built on a thick raft of flint-and-mortar. This timber reinforcement was probably put in because of the anxiety felt about the location of this end of the building. It is interesting to note that the other end of the building, where no intra-mural timbers were employed, has cracked away from the return walls at both sides. The building dates from *c.* 1360. St. Augustine's Abbey at Canterbury was excavated during the early 1930s. One building lying to the south of the infirmary hall contained an intra-mural timber chase in its east wall. The structure was probably erected during the fifteenth century.

The examples described above demonstrate the use of intra-mural timbers as a common accompaniment of medieval building, used in the various types of building common to the ecclesiastical world—churches, towers and ancillary monastic buildings. The large monastic orders were amongst the greatest builders of the Middle Ages and their buildings can be taken to represent what was most characteristic of architecture at the time. This is true of the largest buildings which represent the finest architecture and of the humbler buildings which represented the commoner, but still good-quality, structures. The use of intra-mural timbers in both kinds of building demonstrates their universality and also the importance their function was considered to have by the designers of the buildings.

The other important branch of architecture in stone during the Middle Ages was that of military architecture. For present purposes it is taken to include the buildings placed inside fortifications in the same way as ecclesiastical architecture has been taken to include ancillary monastic buildings which often served what might be described as lay functions although they stood inside monastic confines. An example is the first-floor hall common to both environments as it was equally in ordinary domestic building.

Although there are few examples to demonstrate the point, one can visualize in the use of reinforcement in fortification a continuity on the Continent from Roman to medieval times, particularly in the Byzantine world where masonry fortifications continued to be built throughout the period. One example from this intervening period is that of the walls of Salonica already referred to. In Britain there is a discontinuity and the technique only appears again in Norman times. Richmond Castle was built in early Norman times, the curtain walls and towers dating from before 1086 (DoE *Guide*). The eastern curtain wall had been raised on a stratum of blue clay which dropped away from the walls of the castle. A stream developed in the same direction, running under the castle wall, with the result that the blue clay became so slippery that part of the east wall and a tower fell. It was during excavation in 1926 to investigate and repair this damage that the evidence for intra-mural timbers came to light. Running through the wall longitudinally were two tiers of timber chases, both of which were linked with chases running at right-angles into and through the tower floor. As far as can be ascertained from a drawing made at the time (fig. 5) the timbers ran along a short length of the wall only, and perhaps were inserted by the builders at this point because of the presence of the blue clay which would have obviously have been unsatisfactory as a subsoil. This would perhaps account for the isolated transverse timbers at each end of the threatened section, and the vertical timber that appears to have linked the two tiers in section C–D. Both features would have tended to prevent lateral movement. Another precaution was the driving of sapwood piles into the blue clay layer to form a base on which the rubble and mortar wall was erected. That these precautions were insufficient is only too clear. The sections appear to show that the longitudinal beams rested on top of the transverse ones, and perhaps were nailed or fixed together with wooden pegs. This suggests that jointing was not always thought desirable, and this supposition is strengthened by the evidence that was obtained from Lewes Priory. Unless the builders had a jointed framework prepared beforehand, it would have been necessary to halt the masons whenever they came upon a doubtful section of subsoil and wait for the carpenters to make a frame before continuing with the construction. This would have been an inconvenience and it would have been easier to put in beams (which one assumes they would have had to hand for other purposes), nail or peg them together, and trust to the mortar setting round them to hold them in place.

At Castle Acre Castle in Norfolk, built at about the same time as Richmond Castle, the outer bailey wall is built on a bank overlooking the river. Part of the wall at this south side was reinforced with intra-mural timbers, the empty chases of which remain today. It is possible that the technique was used here because of the location of the wall on top of a made bank which takes a long time to settle and which is always liable to slump.

In Domesday Book, a certain (?) Waldin is mentioned as engineer of Lincoln Castle in 1086 (Harvey, J., 1954). Part of the existing curtain wall is probably his work, and he erected it upon a rough timber framework on top of a made bank. The framework comprised three or four parallel lines of beams laid upon rubble masonry. The longitudinal timbers were crossed by transverse timbers at intervals (Willson). This information was obtained during repair work in the mid nineteenth century, but the observers did not note whether the beam chases intersected, or distinctly passed over each other at the intersections, nor did they measure the dimensions of the chases.

Evidence for the use of intra-mural timbers in castle building in France is provided by Brionne Castle (Eure) built soon after 1090 (Caumont). At the base of the walls of the keep where they are between 3 and 3.66 m. (9 ft. 10 in. and 12 ft.) thick, there are chases which contained horizontal beams sunk in the masonry. A similar example is provided by the castle at Lewes in Sussex which has two mounds. One of these is crowned by an elliptical shell-keep built in the late eleventh century or early twelfth century (Godfrey). This is one of the earliest shell-keeps in the country. At one point in the circumference part of the wall has broken away and slipped down the side of the mound. In the cross-section thus created in the standing wall, one can see the end of a beam chase running longitudinally into the wall. The line of the beam can be traced where the chase has been interrupted by later recesses cut into the masonry and it is possible to estimate the length of the beam as *c.*6.5 m. (21 ft. 4 in.) and its scantling

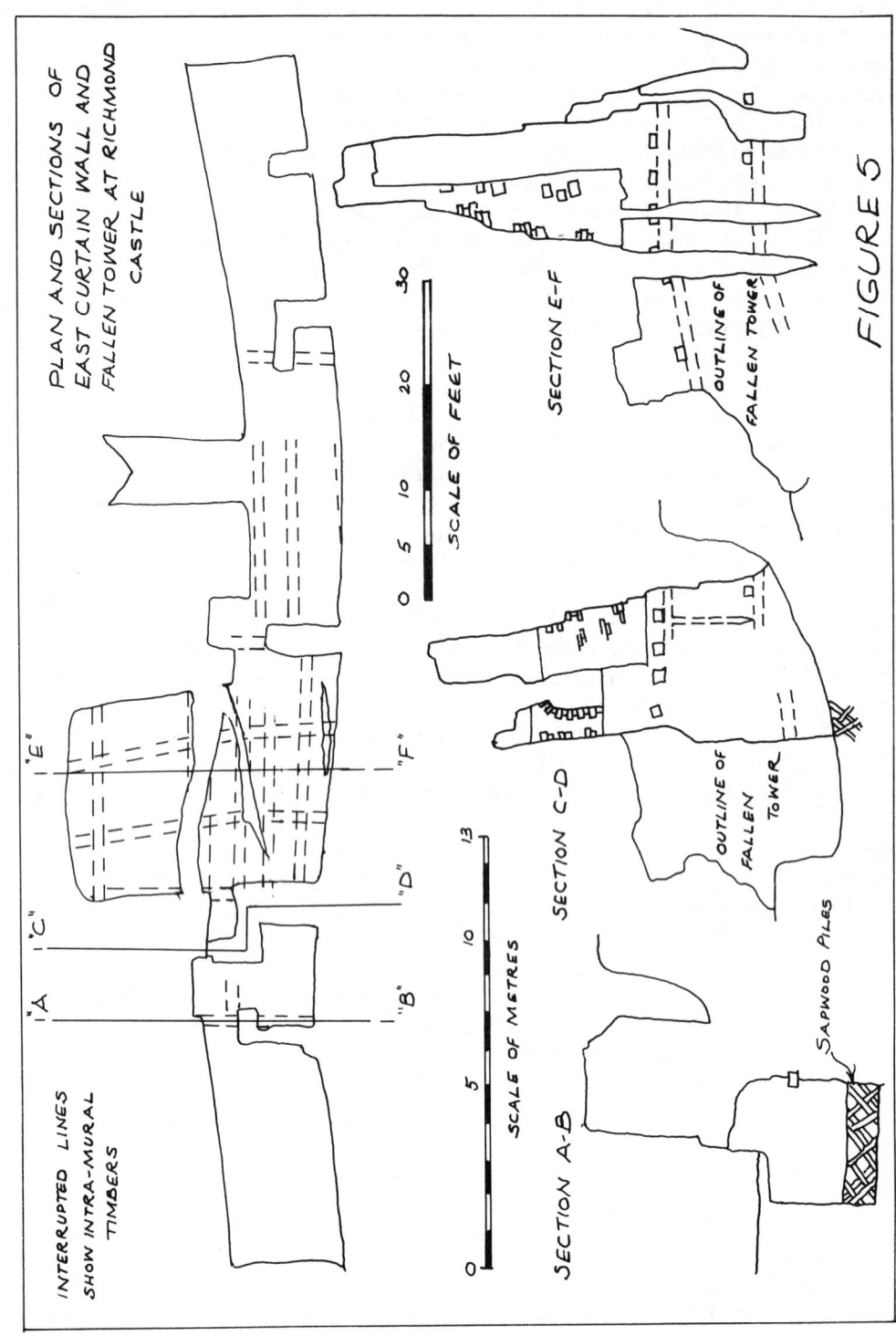

PLAN AND SECTIONS OF EAST CURTAIN WALL AND FALLEN TOWER AT RICHMOND CASTLE

INTERRUPTED LINES SHOW INTRA-MURAL TIMBERS

SCALE OF FEET
0 5 10 20 30

SCALE OF METRES
0 5 10 13

SECTION A-B

SECTION C-D

SECTION E-F

OUTLINE OF FALLEN TOWER

OUTLINE OF FALLEN TOWER

SAPWOOD PILES

FIGURE 5

dimensions as 25 by 17.5 cm. (10 by 7 in.). There is no evidence, in the present remains of the castle, for any other intra-mural timbers, and it appears that the builders' anxiety about the stability of the wall where they inserted the beam was well founded. At Eynsford Castle in Kent, the curtain wall is broken at one point where a gatehouse stood. In the broken walls on either side can be seen the chases once occupied by intra-mural timbers. They ran longitudinally through the wall, reinforcing at a level over a metre from ground level. The castle was built *c.*1100 (DoE *Guide*).

In 1126, William de Corbeuil, Archbishop of Canterbury, built the great keep at Rochester Castle (DoE *Guide*). It was reinforced extensively with intra-mural timbers into which the floor joists were jointed. The timbers ran longitudinally through the cross-wall and the northern wall and probably through the southern one as well at each floor level. They measure 12/15 cm. (5/6 in.) by 10 cm. (4 in.) in scantling and were jointed into much larger floor joists measuring 31 by 25 cm. (12 by 10 in.). This type of construction is similar to that at Bridgnorth (see below) where the same pre-determined arrangement of floor joists and intra-mural timbers was used.

Plympton Castle in Devon is a shell-keep standing on a mound. Apparently it was damaged beyond repair by Stephen *c.*1135 and there is no record of rebuilding. If this is so, then it is likely that the structure belongs to the first third of the twelfth century (Rowe). About two-thirds of the circumference of the keep remain and running throughout the wall are two tiers of intra-mural timbers, one near the present ground level and the other some 2 m. (6 ft. 6 in.) higher up. The scantling dimensions of the beams used was approximately 41 by 25 cm. (17 by 10 in.), with the wider sides forming the tops and bottoms of the chases. The beams do not appear to have been jointed together at their ends, for the casts left behind them in the flint and mortar show that they varied in level. This does not appear to have troubled the builders for the bonding function would not have been substantially impaired by the short gaps left between them. At Merdon Castle in Hampshire there remain above the ground only a few masonry fragments. One of these, supposed to be part of a gatehouse, contains in one broken wall the remains of intra-mural timber chases. One slot is complete but masonry decay has left only traces of one side of the other. The beam slots are about 2 m. (6 ft. 6 in.) apart and there are no signs of transverse timbers. Merdon Castle was built by Henry de Blois, Bishop of Winchester, in 1138 (*Archaeological Intelligence*).

New Buckenham in Norfolk contains a shell-keep built against the side of a ringwork so that part of the circumference of the keep appears to be bonded into the bank behind it (fig. 6). The shell-keep is divided into two by a cross-wall and one enters the northern half through the only door. From this entrance, in a southerly direction within the thickness of the wall, runs a single line of intra-mural timber chases. These continue as far as the junction of the cross-wall with the north-eastern half of the building. At intervals, intersecting with these chases, transverse chases radiate from the inner face of the wall through the masonry and may continue into the rampart behind, thus originally bonding the keep into the ringwork. However, this suggestion cannot be confirmed without excavation. The castle may date from between 1140 and 1150 (Renn). Castle Rising in the same county has a four-sided keep standing inside an extensive ringwork in a central position. The keep is divided into two by a cross-wall which is reinforced by

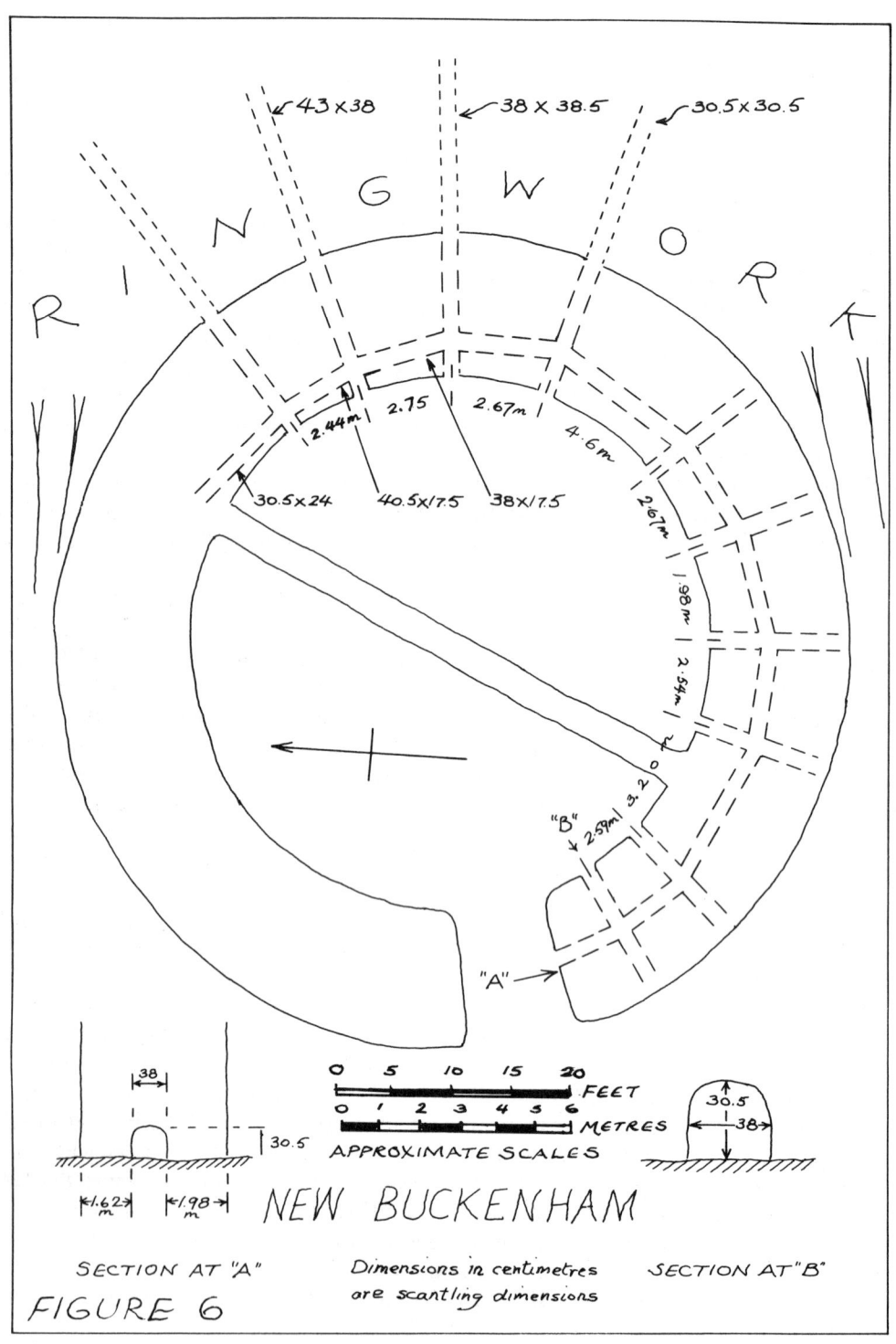

RINGWORK

43 × 38   38 × 38.5   30.5 × 30.5

2.44m   2.75   2.67m   4.6m

30.5×24   40.5×17.5   38×17.5   2.67m

1.98 m

2.54m

"B"   2.59m   3.2 0

"A"

0   5   10   15   20
FEET
0   1   2   3   4   5   6
METRES
APPROXIMATE SCALES

38

30.5

38

30.5

1.62 m   1.98 m

NEW BUCKENHAM

SECTION AT "A"   Dimensions in centimetres   SECTION AT "B"
are scantling dimensions

FIGURE 6

intra-mural timbers at first-floor and roof level. The building dates from the mid twelfth century (DoE *Guide*). Further south in East Anglia, Framlingham Castle stands today as an impressive curtain wall enclosing an area once containing the living quarters of the castle garrison. One of these buildings was a hall which survived from an earlier castle on the site. This hall was incorporated into the complex contained within the new curtain wall and, although the hall is now demolished, it is still represented by one end wall standing embedded in the curtain. In this remnant there remain intra-mural timber chases running horizontally and interested at intervals by joist holes which originally ran at right-angles to it. The hall was probably built between 1150 and 1160 (DoE *Guide*).

The remaining fragment still above ground at Ludgershall Castle is a tower built on the edge of the inner ditch. The wall at the lip of the ditch has fallen, so now the monument consists of three walls containing intra-mural timber chases, in a tier about 2 m. (6 ft. 6 in.) above ground level. (pl.2) The dimensions of the tower and the plan of the timbers are shown in fig. 7. Excavations[3] indicate that the tower was built in the latter half of the twelfth century. In the missing north wall (which was found on excavation to be lying face down in the inner ditch), there would have been insufficient thickness for more than a single chase, and it is so shown in the reconstructed drawing, with a single chase also around the corner latrine tower. Although the timbers were arranged somewhat haphazardly, they do conform to a recognizable pattern and were an integral part of the construction of the building which itself was not planned or built to the highest standards of workmanship. At one point in the south wall (marked "A" on the plan) there is evidence of jointing with a timber at right-angles to the tower wall (fig. 7). This timber may have passed longitudinally through the wall of another building that was aligned N.W./S.E. from the present tower and was bonded into it in this way.

Bridgnorth, Shropshire (fig. 8), is another tower in which extensive use of intra-mural timbers was made. They were arranged in double tiers (pl.3) and ran through the walls of the tower at the level of each floor. Professor E. M. Jope, in his unpublished study of the timberwork, describes the three tiers of timbers, each running through the four walls with the floor joists jointed into them to form a rigid framework at each level. This system provided three floors and an attic. Analogies to this system include Rochester Castle (see above) and the castle at Coucy-le-Château (Aisne) in France (see below). These three buildings provide the best evidence for the combination of floor joists and intra-mural timbers in medieval building. At Bridgnorth Castle the second-floor joist holes in the north wall can be seen aligned in the same plane as the intra-mural timber chases visible in the east (right-hand) wall (pl.4) At the first-floor level, intra-mural timber chases are visible in the east wall, but the floor joists ran at right-angles to those of the floor above—they did not run into the north wall. Professor Jope dates the tower to 1170 on documentary evidence.

Two Welsh castles, both in Powys, present evidence of intra-mural timbers. They are Bronllys Castle built before 1150 (Clark) and Tretower Castle built between 1150 and 1175 (Clark). At Bronllys, the broken walls of the tower show a single timber chase running longitudinally through the circumference near the base of the wall. The chase is almost square in section but there are no dimensions available and it is not clear whether the chase completely encircles the tower. Tretower is a rare example of a Norman keep that has been gutted and the central

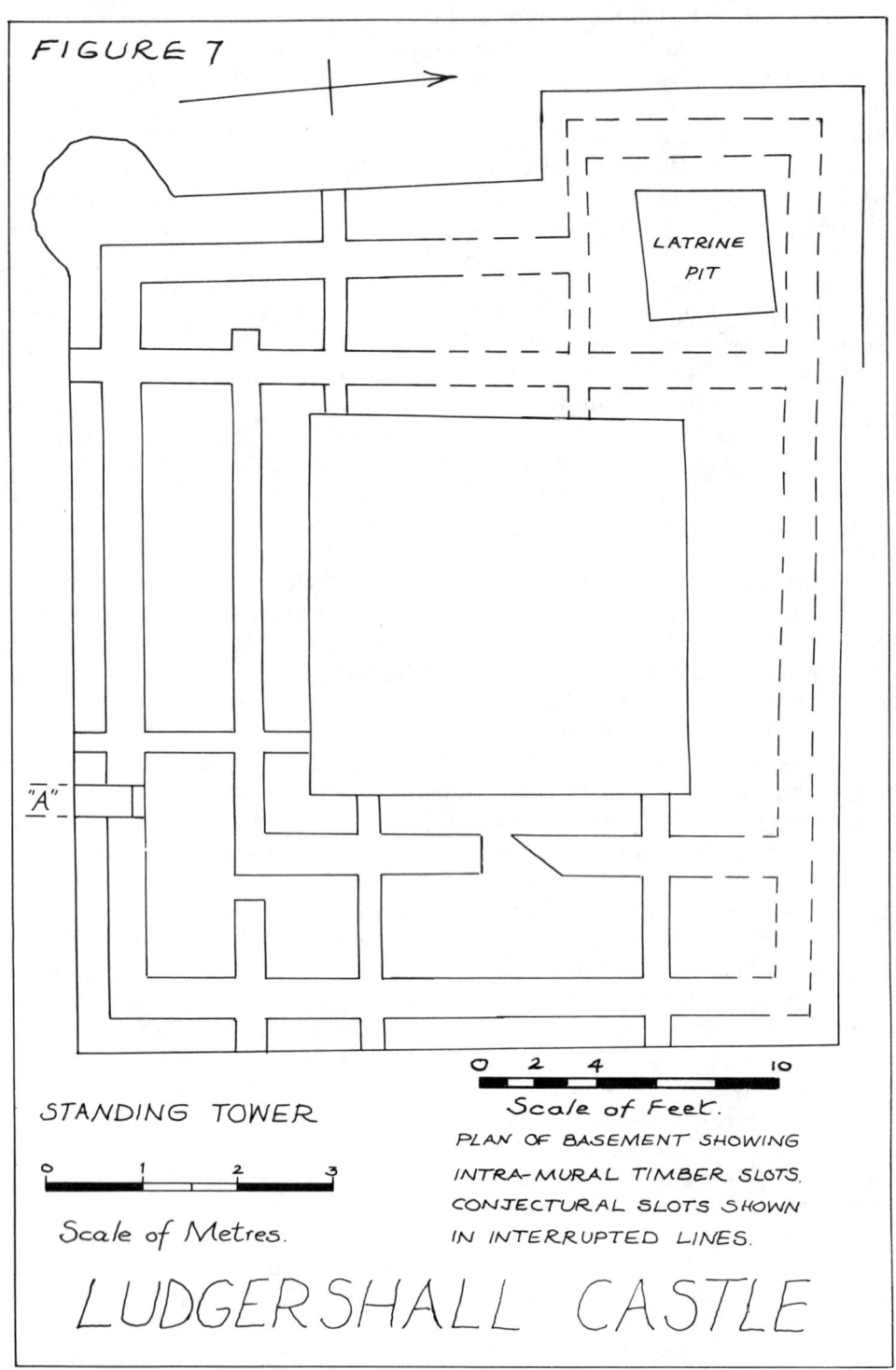

FIGURE 7

LATRINE
PIT

"A"

STANDING TOWER

0   1   2   3
Scale of Metres.

0   2   4   10
Scale of Feet.

PLAN OF BASEMENT SHOWING
INTRA-MURAL TIMBER SLOTS.
CONJECTURAL SLOTS SHOWN
IN INTERRUPTED LINES.

LUDGERSHALL CASTLE

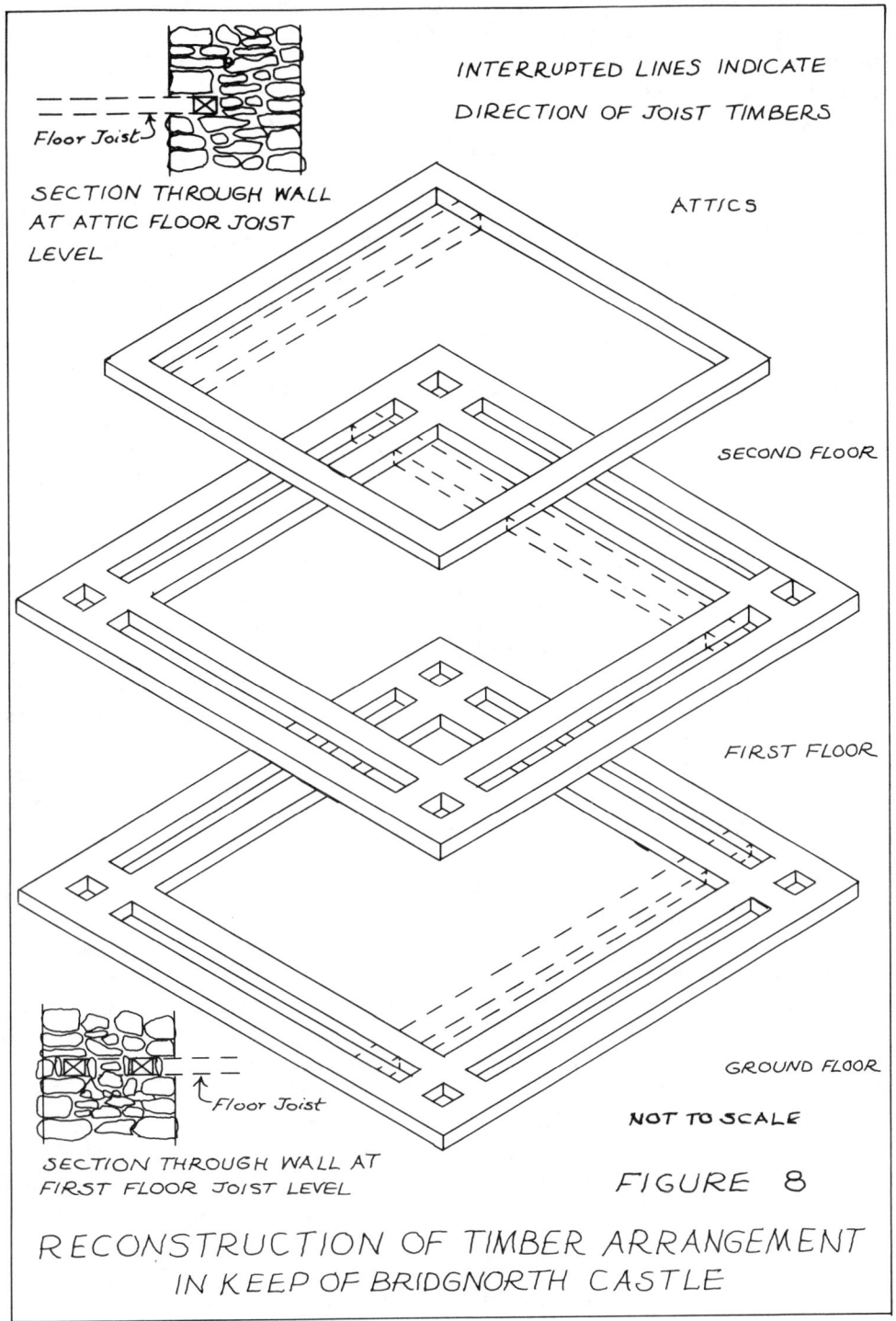

SECTION THROUGH WALL
AT ATTIC FLOOR JOIST
LEVEL

Floor Joist

INTERRUPTED LINES INDICATE
DIRECTION OF JOIST TIMBERS

ATTICS

SECOND FLOOR

FIRST FLOOR

GROUND FLOOR

NOT TO SCALE

FIGURE 8

Floor Joist

SECTION THROUGH WALL AT
FIRST FLOOR JOIST LEVEL

RECONSTRUCTION OF TIMBER ARRANGEMENT
IN KEEP OF BRIDGNORTH CASTLE

part occupied by a later round tower. The intra-mural timber chase is in the Norman keep and is in much the same position as the chase at Bronllys, being low down in the wall. It is also square, about 35 cm. (14 in.) scantling, and here again it is not clear whether the timber encircled the structure which was built on a tump of gravel rising about four metres (13 ft.) out of a marshy meadow.

At Old Sarum in Wiltshire, intra-mural timber chases appear in a fragment of curtain wall north of the gatehouse. This section of wall still stands to a height of 4 m. (13 ft.) and is 3 m. (9 ft. 9 in.) thick. It was reinforced by a double tier of intra-mural timbers, the beams being about 2.45 m. (8 ft.) above the present ground level and about 1.85 m. (6 ft.) apart set in a flint and chalk rubble wall with sarsen bonding. Double tiers like this appear at Hay-on-Wye (see below), Bridgnorth and Ludgershall.

At the end of the twelfth century a number of French castles were built which contained intra-mural timbers. The Castle of Chatillon-sur-Seine (Côte-d'Or) built before 1184 (Enlart) contained bonding timbers that tied one of the towers into the curtain wall. Two timbers were used, one running through the tower wall and the other continuing along inside the curtain wall. Presumably, they were half-jointed at their intersection, for they are in the same plane, but the wall has been broken away at that point.

At Gisors (Eure), the castle was rebuilt by Henry II of England between 1161 and 1184, and the rebuilding included the curtain wall which was reinforced by bonding timbers (Enlart). In the same way, the curtain wall at Moncontour (Vienne) was strengthened in the second half of the twelfth century. Timbers were used in the wall of the keep at Malesmains at Vieuxborg (Calvados) in the later twelfth century, while a similar use was made of them at St. Verain (Nièvre) (Caumont). The castle of Château-Gaillard (Eure) built by Richard Coeur de Lion on his return from the Third Crusade in 1196 was placed on a chalk bluff overlooking the River Seine. At one point in the circuit of the curtain wall of the upper bailey the masonry was bonded into the natural chalk by an intra-mural timber. Later weathering of the surface of the chalk has exposed the now empty chase.

With the beginning of the new century, it is necessary to return to Britain and Ireland for further examples of the use of intra-mural timbers. In Powys again, the small town of Hay-on-Wye has a castle whose remains are principally those of the structure rebuilt by Henry III in 1231. A stretch of the curtain wall is bonded with a double tier of intra-mural timbers. At Rinn Duin in Roscommon in southern Ireland, the castle dates from the thirteenth century. It is an enclosure with buildings each side of the gateway. In the outer wall of one of these are intra-mural timbers at least 3.5 m. (11 ft. 6 in.) long and approximately 30 cm. (12 in.) square.

Finally, at Coucy-le-Château near Laon (Aisne) the keep of the castle surpassed Bridgnorth in the completeness of its intra-mural timber plan before its destruction during the First World War. The joists were tied into the intra-mural timbers at each floor in the same way as they were at Bridgnorth, but the tower at Coucy was circular and the floor joists, encased in mortar, radiated from the centres of the floors to join with the 25 by 30 cm. (10 by 12 in.) intra-mural timbers. The three floors were each 30.75 m. (100 ft.) in diameter and, together with a basement, made a tower 61.5 m. (131 ft. 8 in.) tall which was completed *c*.1240 (Clark).

Returning to Britain, the castle of Ewloe in Clwyd contains a lower ward probably built by Llewelyn ap Gruffyd about 1257 (DoE *Guide*). One of the prominent parts of this structure is the round west tower which was bonded with intra-mural timbers. Another tower bonded with intra-mural timbers is the later fourteenth-century Threave Castle in Kirkcudbrightshire. The timbers formed a massive collar inside the walls at third-floor level: the empty chases measure 50 by 40 cm. (20 by 16 in.). Originally the timbers must have passed through the window openings before the window frames were inserted, so that the function of this collar was purely temporary and was only expected to operate during the period when the building was settling and the mortar was beginning to harden. After this, the timbers could be cut out and the windows put into use. The last example is in the castle or fortified manor house at Weeting in Norfolk built in the fourteenth or fifteenth century. The tower was reinforced at a high level with timber on at least two sides with a junction at one corner. Possibly, timbers existed in the other two sides so that a collar held the tower together.

Foundation Reinforcement

'The general practice of building mud-brick walls on stone foundations and reinforcing both with timber is so widespread in Syria and the Aegean as well as all parts of Anatolia that it can hardly be used as a criterion of particular comparison' (Lloyd and Mellaart (1956), 121)

With this assurance of the presence of foundation timbers in early building in the prehistoric eastern Mediterranean it is possible to dispense with a catalogue of examples there and instead turn our attention directly to instances of the use of the technique in western Europe.

Several of the Roman forts of the Saxon Shore in Britain have foundations reinforced in this way. Most of them were built between A.D. 286 and A.D. 296. One of the earliest in the series is Burgh Castle in Suffolk (Wheeler, 1932). At one point along the outer face of the north wall, the ends of two transverse grooves, 1.5 m. (5 ft.) apart and measuring 25 by 22.5 cm. (10 by 9 in.) in cross section, can be seen where the soil has been denuded from the base of the wall. Also, the impression of timber-framing can be seen on the underside of a fallen bastion on the south side of the fort (fig. 9, top). Limited excavation carried out at Bradwell-on-Sea in Essex (V.C.H. *Essex*) revealed timbers running along both sides of the wall foundation (fig. 14). Not enough of the wall was examined to enable the excavator to decide whether transverse timbers were part of the design. Pevensey Castle was investigated at different times by both J. P. Bushe-Fox and L. F. Salzman who found cavities in the foundations nearly 27.5 cm. (11 in.) square. This evidence was corroborated by Bushe-Fox. His excavation was undertaken to consolidate the northern wall whose lower courses had been seriously denuded. In his report he says:

'When inspecting the walls before the work was begun, I noticed a series of rectangular grooves in the under surface of the masonry. Fortunately, at one point, it was possible to crawl beneath the wall for a distance of 13 ft. [4 m.], approximately its full width, and I was able to examine these grooves in detail

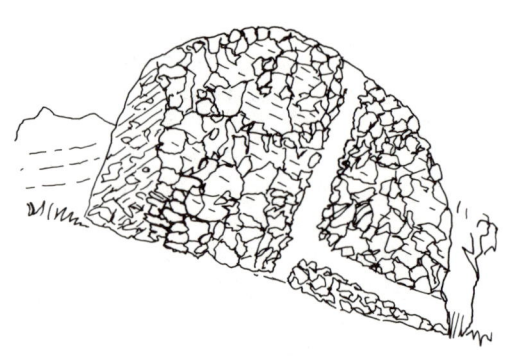

UNDERSIDE OF FALLEN BASTION AT BURGH
CASTLE SHOWING IMPRESSIONS OF FRAMING

FIGURE 9

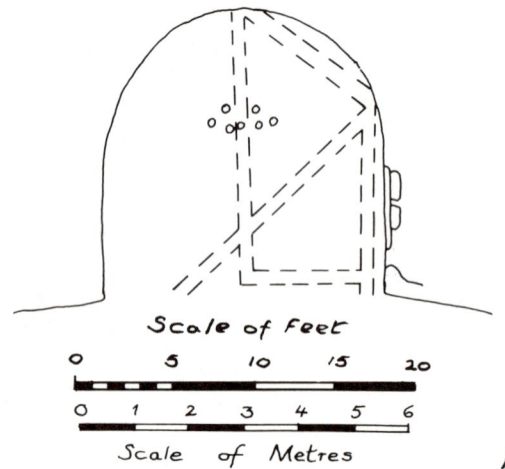

Scale of Feet

Scale of Metres

FIGURE 10

PLAN OF TIMBER FRAMING UNDER BASTION
"B" AT PEVENSEY CASTLE

FOUNDATION TIMBERS ~ BASTIONS

and obtain the plan [fig. 11, plan at C]. The grooves were not confined to one section, but appeared to have extended the entire width of the north side; wherever the lower portion of the outer face was exposed, the ends of the transverse grooves could be seen at regular intervals. There can, I think, be little doubt that they were continuous over the whole circuit, since a similar series of grooves was found on the south side when the berm of the mediaeval ditch at the south-west angle of the inner ward was being cleared. The Roman wall at this point had stood on the crest of a steep slope but, owing to a subsidence, had sheared off from its foundation and now lies in broken sections on the edge of the marsh below. The impressions of the beams [fig. 11, plan at D] in the surface of the chalk-and-flint foundation could be clearly seen but were distorted by the slipping of the soil at the time of the subsidence.' (Bushe-Fox, 60)

Similar grooves were found under bastions to east and west of the fallen portion of the northern wall (figs. 10 and 12). Bushe-Fox goes on to describe the method of construction of the walls:

'A trench was first dug and filled with chalk and flints, and the beams, which appear to have been framed together, were laid upon the surface of this, the space between them being packed with chalk. The masonry of the wall was then built upon this foundation'. (Bushe-Fox, 62)

Bushe-Fox's explanation of the purpose of these timbers is that they were used to consolidate the surface of the packing of flint and chalk by preventing it from spreading out at the sides.

The evidence for the existence of foundation timbers at Portchester Castle was obtained initially during an Office of Works excavation before the Second World War. The footings of one of the Roman bastions (No. 2) destroyed in 1790 were being cleared and this revealed a rectangular foundation with a circular hole in the centre and slots for horizontal bonding timbers (fig. 13). Further evidence for foundation timbers was demonstrated by Professor Cunliffe's examination of the south postern gate. He found that the base of the flint wall had been interlaced with a framework of timbers of about 0.375 m. (15 in.) scantling (Cunliffe, 1963).

Most of the Saxon Shore forts were built in marshy situations near the estuaries of rivers or inlets of the sea. Burgh was built by a tidal estuary into which the present rivers Bure, Yare and Waveney flowed. Bradwell is on the Blackwater estuary while, during the Roman period, Pevensey Marsh was an inlet of the sea and the fort was built at the north-western corner of this haven. Similarly, Portchester commanded the harbour at Portsmouth. These sites presented constructional problems which the builders at Pevensey overcame by selecting a clay hillock as a site. The edges of this island in marshy ground were cut away to expose a flat bed some fifteen feet broad which was prepared for the foundations by driving in oaken stakes. Clay and flint footings were then laid down with foundation timbers on top and the spaces between the beams packed with chalk as Bushe-Fox describes. The same type of construction can be assumed from the section of the wall at Bradwell-on-Sea (fig. 14) although the excavator gives no details of the materials used. It seems reasonable to assume that the foundations at Portchester were similarly built with the timbers performing the same function of preventing the foundations from spreading into the soft ground under the weight

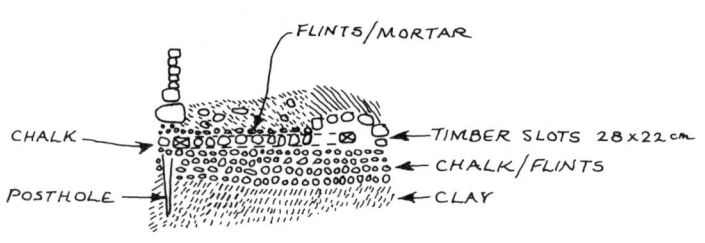

CORE OF WALL

FLINTS/MORTAR

CHALK

POSTHOLE

TIMBER SLOTS 28×22cm

CHALK/FLINTS

CLAY

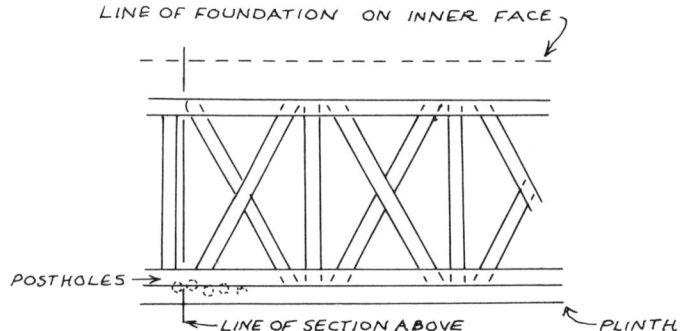

LINE OF FOUNDATION ON INNER FACE

POSTHOLES

LINE OF SECTION ABOVE

PLINTH

PLAN AND SECTION UNDER WALL AT "C"

DISTORTION IN TIMBER FRAMING AT "D"

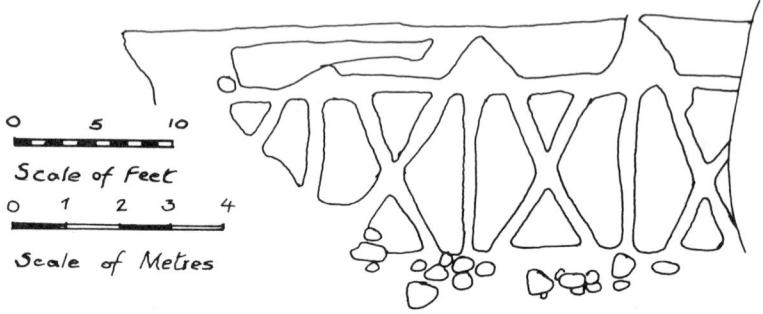

0   5   10
Scale of Feet

0   1   2   3   4
Scale of Metres

REFERENCES ARE TO PLAN DRAWING FIGURE

FOUNDATION TIMBERS ~ PEVENSEY

FIGURE 11

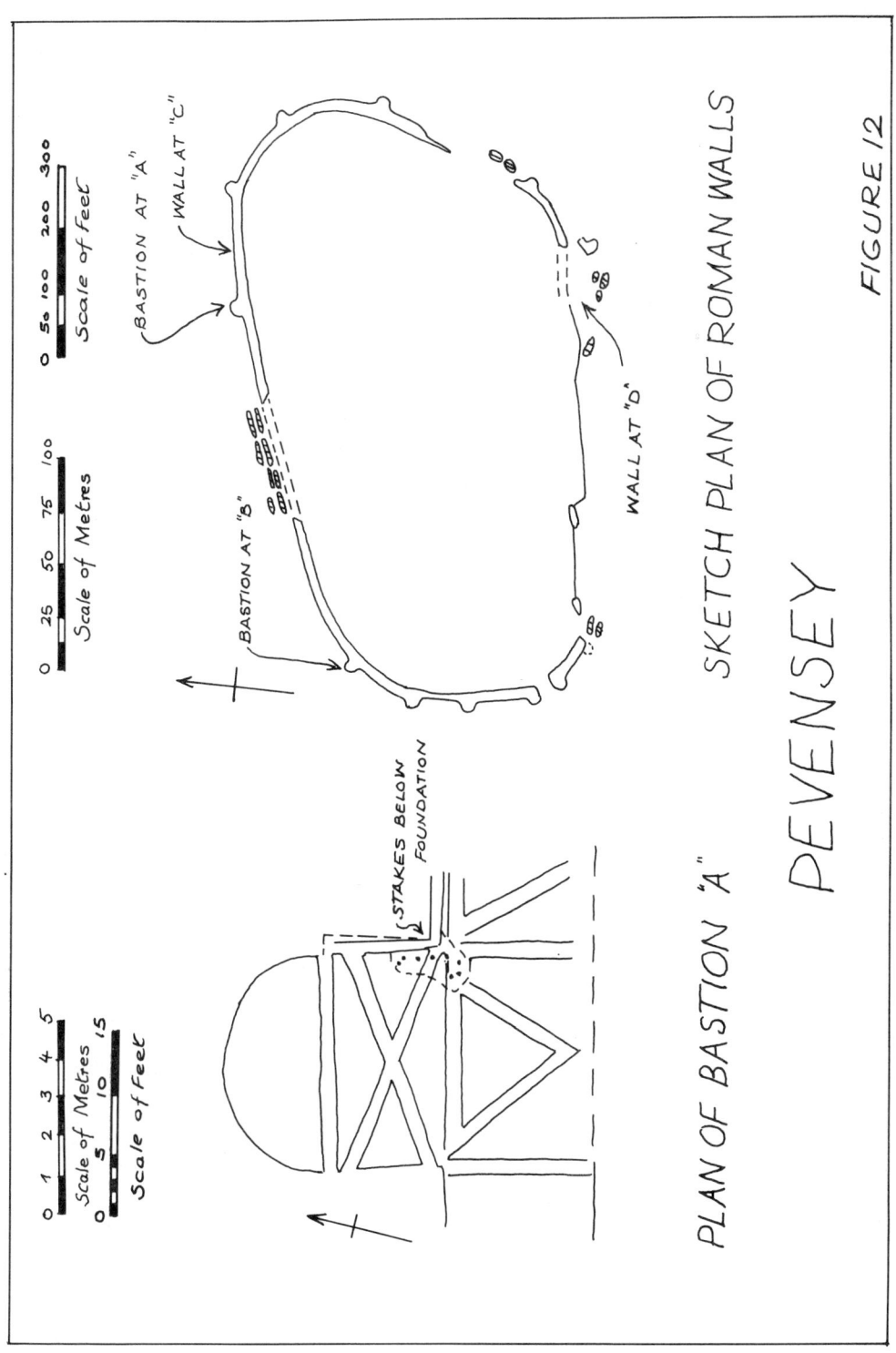

BASTION AT "A"

WALL AT "C"

Scale of Feet
0 50 100 200 300

Scale of Metres
0   25   50   75   100

BASTION AT "B"

WALL AT "D"

STAKES BELOW FOUNDATION

Scale of Metres
0   1   2   3   4   5

Scale of Feet
0   5   10   15

PLAN OF BASTION "A"

SKETCH PLAN OF ROMAN WALLS

PEVENSEY

FIGURE 12

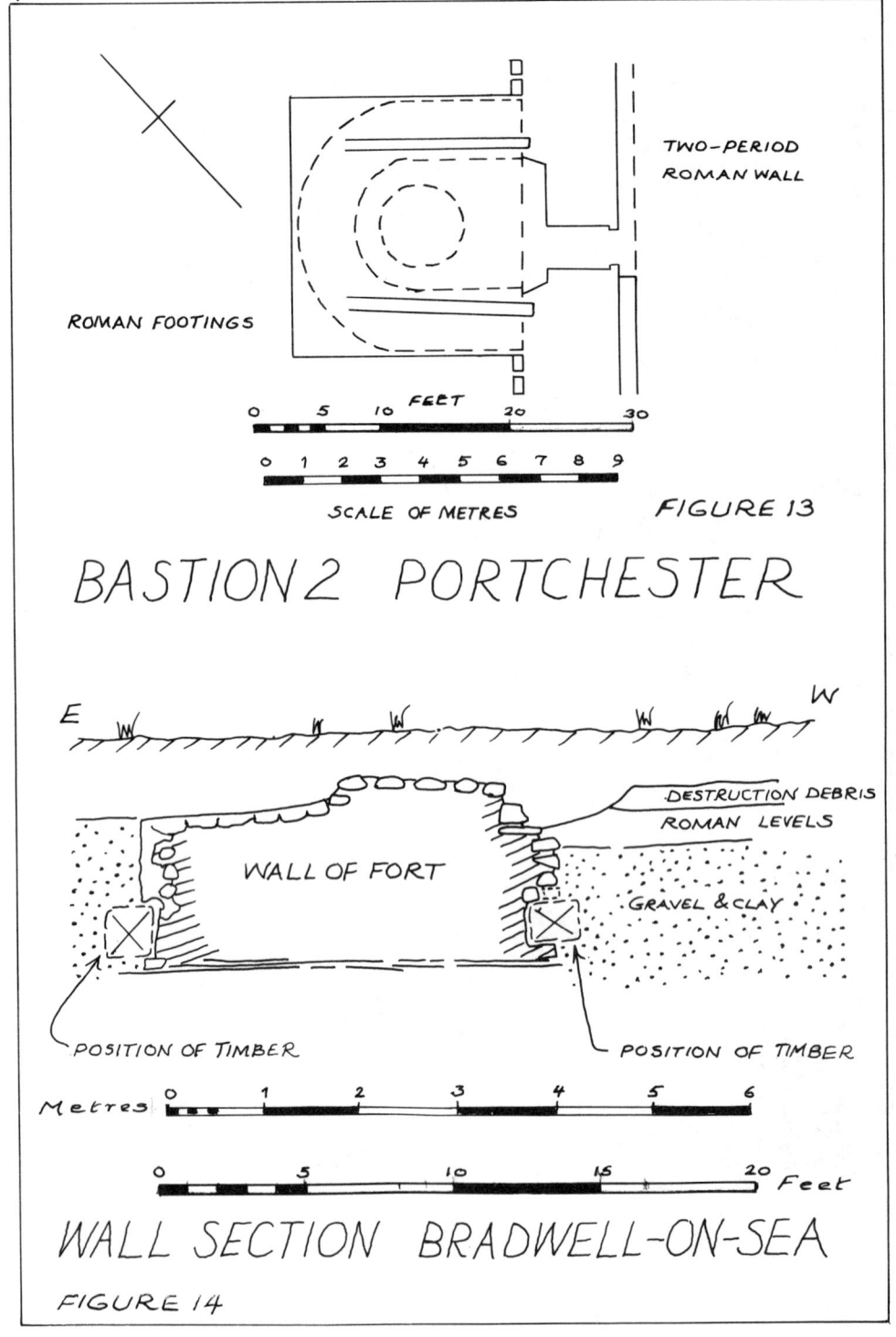

TWO-PERIOD
ROMAN WALL

ROMAN FOOTINGS

FEET
0   5   10   20   30

0 1 2 3 4 5 6 7 8 9
SCALE OF METRES

FIGURE 13

BASTION 2 PORTCHESTER

E   W

DESTRUCTION DEBRIS
ROMAN LEVELS

WALL OF FORT

GRAVEL & CLAY

POSITION OF TIMBER   POSITION OF TIMBER

Metres 0   1   2   3   4   5   6

0   5   10   15   20 Feet

WALL SECTION BRADWELL-ON-SEA

FIGURE 14

of the wall. The longitudinal timbers also prevented the walls from sinking unevenly; they encouraged uniform settling and helped to prevent cracks.

When we move to ecclesiastical building of a later period, we find examples in two major early churches in this country. The tenth-century Old Minster at Winchester has been comprehensively excavated in recent years. In the chalk foundations of one of the bays on the north side of the church were found grooves that once held foundation timbers (pl.5). As the upper part of the chalk foundation had been eroded away and with it the top of the beam cavities, it is impossible to gauge the vertical dimension of the beams, but their width varied between 20 and 30 cm. (8 and 12 in.). The reinforcement consisted of a single line of beams running along the sides and around the apsidal end of the bay to join up at both sides with timbers that ran at right-angles along the foundations of the main body of the church (fig. 15, right). The beams were not set at a uniform depth, but varied within a range of some centimetres. This is brought out by pl.6, showing the western intersection where the through east/west beam was met by the north/south beams which were presumably jointed into it. It is instructive to compare this use of foundation timbers with that employed at York Minster. Sir Charles Peers excavated under the York choir before the First World War. He says in his report:

'The whole [the cathedral foundations] has been strengthened by bond timbers, averaging 12 in. [30 cm.] square, running through the entire length of the foundation, as far as can be ascertained, and tied together at intervals by cross-timbers. Part of these remain, but the majority have rotted away, leaving the chases in the masonry empty' (Peers, 118).

Part of Peers' plan is reproduced in fig. 15, left. As can be seen from the scale, the York apse is larger than the apse at Winchester, and has a double tier of timbers crossed by transverse ties as well as isolated timbers, perhaps interpolated where it was thought that there were places of special weakness. The arrangement is more sophisticated than at Winchester, but probably shows signs of second thoughts in the random placing of the isolated timbers that could well have been incorporated in a slightly more elaborate original plan. Recent excavation at York has given us additional information and corrected Peers' Saxon estimate for their age. The parallel foundation shown in Peers' plan continues westwards to form the foundation of the Norman tower. Similar, but narrower, foundations extend under the nave and transepts. On this evidence, the foundations are Norman and belong to the church that included the tower which was built by Archbishop Thomas of Bayeux *c.*1080. The following is a description of the method of constructing the foundations:

'A large area was cleared of its buildings and excavated 4/6 ft. (1.2/1.8 m.) below ground level. Normally, the whole site was excavated, but sometimes a trench was dug (about 25/30 ft. (7.6/9 m.) wide for the 21-ft. (6.4-m.) wide foundation). Within this excavation, the lines of the foundations were marked out with wooden pegs 2/3 ft. (0.61/0.915 m.) long by circa 4 in. (10 cm.) wide, and then the foundations were built up in layers of mortared, haphazard rubble, circa one foot (0.3 m.) deep between faces of coursed stone. The excavation was then filled up to the level of the top of this layer and the infill served as a platform on which to build the next layer. Within the base layer and the layer next above chases with coursed sides were constructed to take the lateral and longitudinal timbers. The work proceeded in this way until ground level was reached'.[4]

Building on this sort of scale was rare in eleventh-century England and must rank with structures like the White Tower in London as the most advanced of the period.

At Old Sarum in Wiltshire, a building probably used as a treasury and library was built at approximately the same time as the first cathedral on the site which was the work of Bishop Osmund between 1078 and 1092. Excavation in 1915 (DoE *Guide*) revealed that one wall of the treasury had been bonded at foundation level with a double line of beams, 22.5 by 30.5 cm. (9 by 12 in.) and 20 by 19 cm. (8 by 7¾ in.) in scantling, placed side by side in a wall only 0.775 m. (2 ft. 7 in.) wide. This allowed for 7.5 cm. (3 in.) of masonry thickness on one side of the double timber and a 17.5 cm. (7 in.) thickness on the other. Why such an extraordinary technique was used is impossible to say: one would think that such slender skins of masonry would be impracticable. It is perhaps significant that the greater part of the contemporary cathedral was destroyed in a storm five days after its completion.

There are several examples of foundation timbers in military building in Britain. Sheffield Castle, built *c.*1100 (Hunter), when excavated in recent years, was found to be resting, at one point at least, on made ground covered with stone slabs beneath a horizontal beam on oak supports. At Chilham Castle, Kent, a keep was built before 1171 over the foundations of an eleventh-century hall (Clapham). Where it oversailed the earlier walls, the keep was carried on a bressummer of timber which supported the flint rubble walls in the same way as the foundation timbers did at Brough where they rested on late eleventh-century work placed on the foundations of a Roman barrack block. This evidence was obtained in an Office of Works excavation in 1927. The keep above the foundation timbers at Brough was built *c.*1200 (*Cumberland and Westmorland*). Clifford Castle in Herefordshire, built probably in the early thirteenth century, has two examples of foundation timber chases. One is under the curtain wall and the other in the south-eastern corner of the hall. Today there exists only one arm of the corner reinforcement in the south wall; the corner has broken away and the east wall is down so that it is impossible to verify the existence of the east arm of the reinforcement. Also in Herefordshire, the earlier castle at Goodrich was built at about the same time as Clifford (R.C.H.M. *Herefordshire*). The present structure is substantially of the late thirteenth century but there are remains of the early thirteenth century. One of these is the foundations of the south-west tower, which were reinforced with timbers. No further details apart from those given in the plan and account by the Royal Commission are available so that it is impossible to say anything about the arrangement of the timbers. Foundation timber chases can be seen in the base of a blown-up tower in the castle of Corfe. This tower stood west of the gatehouse in the outer curtain wall. It is 9.10 m. (30 ft.) in diameter and solid up to the gorge. Such a massive structure would hardly seem to have needed reinforcement, but it stood on the edge of a slope and medieval builders seem to have been particularly sensitive about structures in such a position. This one was

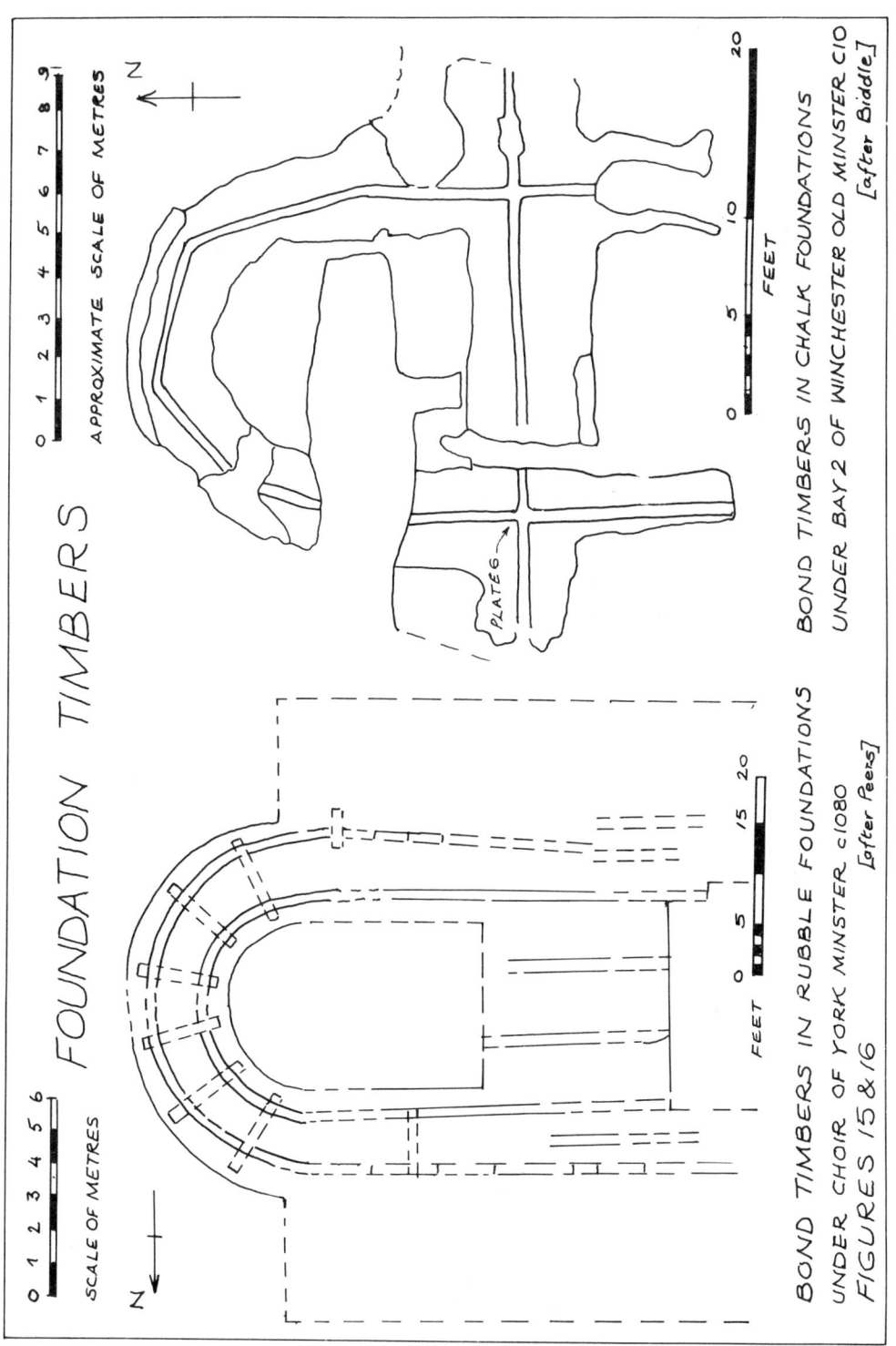

FOUNDATION TIMBERS

BOND TIMBERS IN RUBBLE FOUNDATIONS
UNDER CHOIR OF YORK MINSTER c1080
[after Peers]

BOND TIMBERS IN CHALK FOUNDATIONS
UNDER BAY 2 OF WINCHESTER OLD MINSTER C10
[after Biddle]

FIGURES 15 & 16

APPROXIMATE SCALE OF METRES
0 1 2 3 4 5 6 7 8 9

SCALE OF METRES
0 1 2 3 4 5 6

FEET
0 5 10 20

FEET
0 5 15 20

PLATE 6

bonded with a timber framework: the larger beams 45 by 30 cm. (18 by 12 in.) scantling and others 23 cm. (9 in.) square. The outer curtain wall, with its towers, has been dated to between 1265 and 1269 (Clifford Perks).

Before passing on to the third type of reinforcement, it will be as well to summarize the evidence for the use of intra-mural timbers and foundation timbers during the medieval period. There follows a classification of the examples arranged according to the types or the parts of the building in which they were used. Some, whose description was omitted in the interests of brevity in the previous pages, are listed and the ecclesiastical sites are marked with an asterisk:

Intra-mural timbers

Towers:	* S. Ambrogio, Milan	C9
	* S. Carpofore, Como	1040
	Brionne Castle (Eure)	after 1090
	* S. Abondio, Como	1095
	* Baugy church, Burgundy	C11
	* S. Marco, Venice	1063–1111
	Rochester Castle	1127
	Bronllys Castle, Powys	before 1150
	Castle Rising Castle	mid C12
	* St. Denis (Seine)	mid C12
	St. Verain Castle (Nièvre)	mid C12
	Ludgershall Castle, Wilts.	?1160
	Bridgnorth Castle	1170
	Malesmains Castle (Calvados)	late C12
	Castle of Coucy-le-Château (Aisne)	*c.*1240
	Threave Castle, Kirkcudbrightshire	1369–*c.*1400
Gateways:	* St. Ethelbert's Gate, Norwich	1090
	Merdon Castle, Hants.	1130

Rectangular buildings (not keeps):

	* Sv. Ohrid	852–89
	* Ala Klisse, Anatolia	C9
	* Old Sarum library/treasury	1072–92
	* Lewes refectory	1090–1100
	* La Madeleine, Vézelay (Yonne)	1132
	Wolvesey Hall, Winchester	1129–35
	* Reredorter, Lewes Priory	1140–50
	Framlingham Castle, Norfolk	1150–60
	Clifford Castle, Herefordshire	early C13
	* Derry church, Co. Down	C13
	* Maghera church, Co. Down	C13
	* Wilmington Priory, Sussex	C13
	* Kilmacdaugh monastic building, Co. Galway[5]	C14/C15
	* Castle Acre Priory, Norfolk	*c.*1360
	* St. Augustine's Abbey, Canterbury	?C15

Shell keeps:

Lewes Castle, Sussex	*c.*1100
Plympton Castle, Devon	early C12
New Buckenham Castle, Norfolk	1140–50
Tretower Castle, Powys	1150–75

Mural towers and walls:

Salonica town walls	*c.*415
Richmond Castle, Yorkshire	1066–86
Castle Acre Castle, Norfolk	before 1086
Eynsford Castle, Kent	*c.*1100
Der Husterknupp, Niederrhein, Germany	*c.*1150
Gisors Castle (Eure)	1161–84
Château-Gaillard (Eure)	*c.*1196
Old Sarum Castle, Wiltshire	late C12
Moncontour Castle (Vienne)	late C12
Chatillon-sur-Seine (Côte d'Or)	late C12
Rinn Duin Castle, Roscommon[6]	C13
Hay-on-Wye Castle, Powys	1231
Ewloe Castle, Clwyd	1257
Corfe Castle, Dorset	1265–69

Foundation reinforcement

* Winchester Old Minster	C10
* York Minster	1080
Richmond Castle, Yorkshire	1066–86
Lincoln Castle	1086
Sheffield Castle	?*c.*1100
Goodrich Castle, Herefordshire	1160–70
Chilham Castle, Kent	before 1171
Brough Castle, Cumbria	*c.*1200
Clifford Castle, Herefordshire	early C13
* Westminster Abbey nave	1475

The examples listed above span the years from the fifth to the fifteenth century, but the techniques do continue in use and there are later examples at Hillsborough, Co. Down (1650–60) (Jope, 1966) and in domestic building in England in the nineteenth century when it was an accepted method for bonding bricks as well as for reinforcing footings (*Dictionary*), despite an attempt by Brunel to introduce a system of bonding by the use of hooped iron. It was really only ousted by cast-iron beams. But this was only in thick walls: in thin walls, timber was considered superior to iron bond in dwelling-houses because

'it tends to keep walls steady and true while being carried up, and prevents them from being racked by the wind or by the scaffolding' (*Dictionary*)

To a certain extent, the medieval functions of timber reinforcement can be classified according to function. Their use as foundation timbers in different kinds of footings—in chalk at Winchester Old Minster and in rubble at York Minster—was accepted as a natural technique in the earliest masonry large-scale buildings in England. It developed several variations. Winchester had a single line of longitudinal beams crossed by transverse timbers; York had double longitudinal beams and transverse ones; Wolvesey Hall in Winchester had a single longitudinal beam placed just above the footings, and Westminster Abbey nave (Rackham) had crossed beams like those used at Pevensey Roman fort.

Marshy ground seems to have been the usual situation which called for a framework of crossed timbers[7]; ground untrustworthy for other reasons seems to have required the use of longitudinal beams, with or without transverse timbers. Such ground could be on the edge of a slope. Clifford Castle has part of its curtain wall foundations reinforced with timbers and this is because the wall stands on the edge of a steep slope down to the River Wye. In the same way, Lincoln curtain wall was built on timbers on the edge of a slope into a defensive ditch, while Corfe Castle had at least part of its outer defences stabilized in the same way as a precaution against uneven subsidence during the post-construction 'settling' period. At Brough Castle and Chilham Castle, where a later structure was built upon the foundations of an earlier one and the builders were uneasy about the superimposition, they relieved their apprehension by sandwiching a timber bressummer between the earlier and the later work. This framework supported the later wall where it oversailed the earlier walls and gave homogeneity to a structure that might very easily have developed cracks as it settled, because of its uneven site.

In some cases where the ground was uncertain, the walls were reinforced higher up in the wall proper. Richmond is a good example. The builders used sapwood piles to provide a footing for the masonry but did not put in foundation timbers; instead the walls and the tower were laced with intra-mural timbers, both horizontal and, it appears, vertical in one instance. At Lewes, the reredorter was placed over a diverted stream as well as on a slope, and this must have motivated the builders to insert intra-mural timbers into walls that are rather slender (by Norman standards) to support an upper storey. Similarly, where a wall was built close to the edge of a slope, intra-mural timbers were sometimes inserted as at Old Sarum or Castle Acre Castle.

Shell-keeps built on mounds and often very close to the edge of mounds must have given some anxieties over stability and prompted the insertion of intra-mural timbers. At Lewes this is the case and, in a far more elaborate manner, at Plympton. At Ewloe, where the west tower is built on a rocky base and, at Tretower, set upon a gravel hillock, the builders used bonding timbers in the same way. New Buckenham, which was constructed in a marshy locality, had transverse timbers as well as longitudinal ones, and is an interesting adaptation of the foundation technique to a use within a wall proper. The walls are over 4 m. (13 ft.) thick and comprise the largest shell-keep in England and it may have been the scale of the building that prompted the unusual plan. Certainly the timbers were very large, scantling sizes of well over 0.3 m. (1 ft.) being common in the dimensions of the chases of both the transverse and longitudinal timbers. The anxiety about the building's stability is also reflected in the bonding into the ringwork. At Château-Gaillard a similar technique bonded the curtain wall into the natural hillside itself, in contrast to New Buckenham where the timber bonding is into an artificial mound.

Intra-mural timbers were also used for bonding two structures together as in the curtain wall and tower at Chatillon-sur-Seine and also at Richmond. In the tower at Ludgershall, evidence for the addition of another building is provided by a significant jointing halfway along one of the longitudinal timbers (see Appendix A).

At Chatillon-sur-Seine, the right-angle combination of two timbers served a double purpose. Besides bonding two structures together, it also served to brace a corner. Examples of this corner-bracing can be seen also at Ala Klisse, in the hall at Clifford, at Castle Acre Priory and at Maghera church in Northern Ireland where, however, one of the two arms of the brace has broken away. At Lewes Priory reredorter, where a right-angle brace seems to have been called for, it was not used. In the south-eastern corner, the longitudinal intra-mural timber chases in the two walls are at different heights and one does not extend as far as the corner.

Beams at wall-plate level in the gables of the churches are found at Maghera and Derry. These were inserted probably to prevent the distortion of the masonry of the gables should the wall underneath settle unevenly. If this did happen, it would be easy to fill the gap underneath the beam until the top of the wall was once again flush with the under-surface of the beam.

Intra-mural timbers were also used as beds into which to socket floor-joists. It may be that this was a device to hold the ends of the joists firmly in place (examples of this use of intra-mural timbers seem to indicate that jointing was used in some cases) and prevent them from warping or shrinking. A warped joist could have a deleterious effect on the flatness of the floor laid over it. This precaution was perhaps most necessary where oaken joists were used. G.F. Innocent in his book on building technique says:

'English oak cannot be thoroughly seasoned, and oak timber which has been fixed in a building for centuries may warp and twist as if it were new when cut up and used again. So English carpenters used it when it was felled; and the durability of their work is the answer to modern objections' (Innocent, 101)

This technique when it was carried out right through the building at every floor level does leave the fullest record of the presence of intra-mural timbers. Good examples at Rochester, Castle Rising Castle, Bridgnorth and Coucy-le-Château belong to a sophisticated class of workmanship while humbler examples are at St. Ethelbert's Gate, Norwich and at Framlingham Castle.

Instances of intra-mural timbers used as collars to hold a building four-square during the initial settling period occur at Threave and the reredorter at Lewes Priory. The Lewes collar was the upper tier of timbers. In both cases the beams originally passed through the window spaces, emphasizing the temporary nature of the device. It is difficult for us, living in a world in which iron tie-beams and girders have been in use for a hundred years, to understand the anxieties of a builder who was putting up a tall tower like Threave or a building with slender walls. He had no way of framing the building in reinforced concrete. His only solution was to make sure that his building had as long as possible for the mortar to harden and the structure to settle with the bonding timbers in position. It may be that the building stood for months or even for years with a timber collar passing through the window openings. It might be asked why it was not always the practice to put the timber collar in above or below the windows so that it did not interfere with the completion of the building. The answer to this is probably that the collar would be placed, by choice, in the plane of greatest weakness where distortion must be prevented at all costs. This is at the highest window level, and the price of having to delay the completion of the building for a considerable period of time was worth paying in exchange for a stable structure and fracture-free masonry. In the towers of Rochester Castle, Castle Rising Castle, Bridgnorth and Coucy-le-Château the intra-mural timbers at each floor level performed the function of a collar as well as the function described above.

Of the other examples quoted little can be said. It is difficult, for example, on present evidence, to assign any specific purpose to the intra-mural timbers at Merdon, the Old Sarum library/treasury beams or the refectory undercroft timbers at Lewes. It may be that there was some anxiety felt for the foundations that is not apparent today, and this is the most likely explanation. Medieval builders seem to have had as much concern for the longitudinal stability of their walls as they did for the transverse stability, which they generally took care of with buttresses. The longitudinal stability was provided for by intra-mural timbers which are so often placed lengthwise in the walls that this might almost form part of their definition. But this would have been a short-term anxiety, for the danger of unequal settlement was only present in the first few years of a building's life. No doubt it took longer for a building like New Buckenham Castle to settle than it does for a modern house, and one would need to multiply the five years usually allotted to the latter's settling period by as much as six to form a comparative estimate for the former. But thirty or even fifty years is well within the lifetime of an oak beam and it is safe to say that the building was well settled and free from danger of cracking long before the intra-mural timbers had rotted away.

Oak was the most common building timber during the Middle Ages, but other timbers were used. At the port of Sandwich in 1463

'viij gret elmes for planckes for the foundacion'

(Exchequer K.R. Accounts)

were brought to the site and cost forty shillings.

Scantling size varies a great deal even within the same building. Lewes Priory contains many examples of this inconsistency: the chases in which the timbers were contained are of the following sizes:

20 by 17.5 cm.	(8 by 7 in.)	32.5 by 15 cm.	(13 by 6 in.)
20 by 16 cm.	(8 by 6½ in.)	30 by 15 cm.	(12 by 6 in.)
21 by 17.5 cm.	(8½ by 7 in.)		
19 by 16 cm.	(7½ by 6¼ in.)	30 by 33.5 cm.	(12 by 13¼ in.)
16 by 13.5 cm.	(6½ by 5½ in.)	27.5 by 30 cm.	(11 by 12 in.)

The larger square timbers, whose dimensions appear in the third group, were used low down in the wall. The smaller square timbers (first group) occur in all other positions in the wall, while the oblong beams (second group) were only used high up in the wall with the shorter dimensions serving as the base on which the beam rested.

At Ludgershall, the scantling dimensions are similar to those of the oblong beams at Lewes:

45 by 15 cm.	(18 by 6 in.)	35 by 17.5 cm.	(14 by 7 in.)
37.5 by 15 cm.	(15 by 6 in.)	45 by 16/17.5 cm	(18 by 6¼/7 in.)
47.5 by 15 cm.	(19 by 6 in.)	41.5 by 17.5 cm.	(16½ by 7 in.)

But at Ludgershall the beams rested on their wider dimensions throughout the building.

The intra-mural timbers at Wolvesey Palace Hall occurred very low down in the wall and were on average 30 by 37.5 cm. (12 by 15 in.). Oblong timbers also occur low down in the wall at New Buckenham (37.5 by 17.5 cm./15 by 7 in.) together with squarer ones (average size 30 by 37.5 cm./12 by 15 in.). Plympton's two tiers of intra-mural timber chases contained beams approximately 43 by 25 cm. (17¼ by 10 in.). Timbers of this size higher up in the wall are unusual; in most cases large timbers were set lower down in the wall, with smaller scantlings further up.

Some of the timbers appear to have been trimmed before use apart from the usual squaring. At New Buckenham, a number of the timber chases have a rounded top when seen in section in the wall-face and the same thing can be seen at Lewes (pl.7). However, the sections further along inside the chases are very often rectangular, and the suggestion is that ends of the beams were sometimes chamfered so that the mortar would grip them more firmly and prevent any longitudinal movement of the beam. But this is an unusual feature; in most cases the beam chases are roughly rectangular throughout their length.

*Iron and timber reinforcement not encased in masonry
(extra-mural reinforcement)*

So that the reader will understand what kind of reinforcement is under discussion in this section it will be as well to start with a description of the main types. The following are the most common:-

1) as a collar around a structure

This is most common in churches between the supports below a dome or a tower. An example can be found in a western church in S. Sepolcro in S. Stefano, Bologna (pl.14). Further east such reinforcement was used in churches in Salonica—Theotokos church (fig. 45), H. Catherine (fig. 46) and also at the Dair al Ahmar, in Egypt, (fig. 39) and Talich (fig. 30). An example of timber reinforcement acting as a collar around a dome itself is shown in fig. 43 in the Theotokos church in Salonica.

2) across the diameters of a dome

Two tie-beams are sometimes set transversely across the springing of a dome at right-angles to each other so that they intersect in the centre as at Angoulême Cathedral (fig. 63) and Studenica (fig. 47).

3) across a vault[8]

The thrust exercised by a barrel vault is continuous along its length and has to be supported along this length by continuous walls. These walls must be of considerable size and thickness so that the weight of the vault does not push them apart (fig. 17). The higher the vault is raised, the more massive the walls must be in order to contain the horizontal component of the vault's lateral thrust. The thickness may be decreased towards the tops of the walls, but it has to be wide enough all the way up to contain the line of pressure within the middle third of this thickness (fig. 18). If not, the vault will spread and push the walls apart. In order to reduce the necessary thickness, some of the thrust can be absorbed at the level of the springing of the vault by tie-beams or tie-rods (fig. 21). These prevent the arch action from pushing the supports apart. As well as making great thickness of wall unnecessary, the use of ties allow walls to be pierced more readily by window openings.

The same problems hold good for a pointed vault, but here the thrusts are canalized along the vaulting ribs so that the walls do not have to be continuous. Piers set at the corners of the vaulting bays, which take up the thrusts channelled to them along the ribs, are subject to the limitations outlined above and, unreinforced, have to be equally thick at ground level. The reinforcement at the springing of the transverse ribs set between the vaulting bays absorbs a good deal of the thrust, but the Gothic architects usually preferred to use flying buttresses that take up the thrust at the same level exterior to the building and carry it to ground level over the roofs of the aisles.

An example of a change from the use of tie-reinforcement to that of flying buttresses occurred in the church of La Madeleine at Vézelay where, in the original nave structure, the tie-rods were fixed immediately above the corbels supporting the transverse arches (fig. 4). The vaults, however, began to spread, so the tie-rods were abandoned and flying buttresses were used instead. At Maria Laach, in the Rhineland, a copy of the Burgundian church, the builders benefited from the earlier experience and trouble was avoided by placing the ends of the tie-rods higher up in the haunches of the vault.

In the later Gothic churches, vaults were more sharply pointed and this complicated the structural problems. In a free-arch system, such as a vault which carries no surcharge and therefore only supports itself, the line of pressure is an inverted catenary curve. To demonstrate this, one should imagine a chain suspended from two points on the same level but separated by less distance than the length of the chain. Each link is an identical unit of weight along the curving loop of the chain and is in tension with its neighbour. Together, these links form a curving line whose axis is in the line of tension in the freely suspended chain (fig. 19). If this loop of chain was considered as being rigidly fixed so that its curve would not deform and if it were then inverted, each link would then be in compression with its neighbours and the axis of the curving line of links would be the line of pressure. This is what happens in an arch, where the voussoirs constitute the 'links' of the inverted catenary curve.

As we have already seen, it is necessary, in order to ensure stability, for the line of pressure to be contained within the middle third of the structure. If this natural law is broken, as it almost inevitably is in a strongly-pointed vault, trouble follows. When the inverted catenary line of pressure (the broken line with arrows in fig. 20) comes too close to the extrados or the intrados of a pointed arch of masonry or

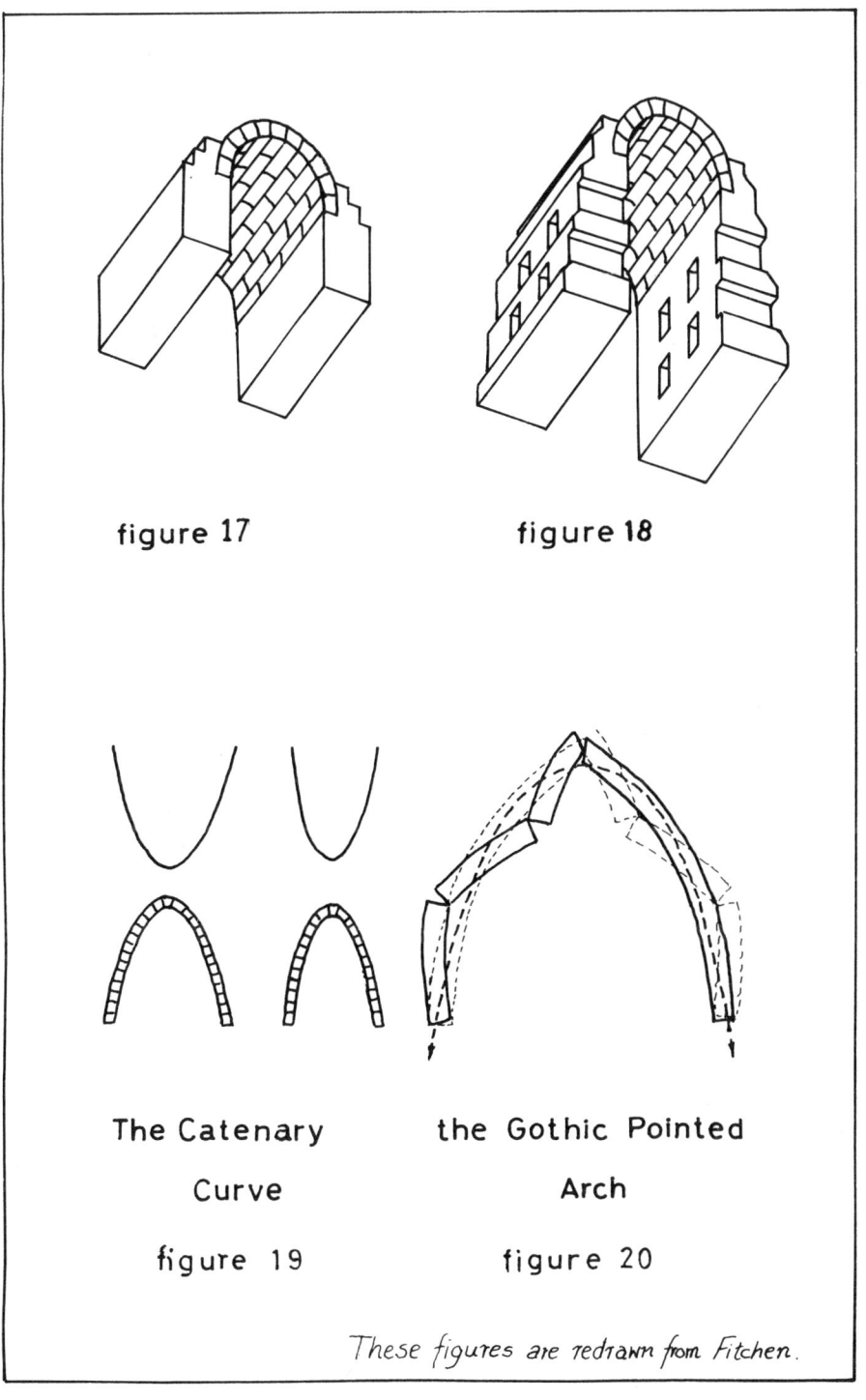

figure 17

figure 18

The Catenary
Curve

figure 19

the Gothic Pointed
Arch

figure 20

These figures are redrawn from Fitchen.

even passes outside the arch (as under the crown), the following consequences ensue (fig. 20):

a) the springs of the arch either spread apart or open their joints at the intrados.
b) the arch bursts outward at the haunches.
c) the joints of the intrados near the crown open.
d) the crown rises.

Clearly, no catenary curve can be pointed, so it usually happens that, in a sharply-pointed vault, the line of pressure passes outside (below) the thickness of the vault crown and perhaps also passes beyond (above) its extrados high along the haunches. In such a situation, the vault would tend to rise at the crown.

But the medieval builders resorted to certain practices to prevent this happening even when their vaults were sharply pointed (fig. 21):

a) they could fill the concave funnel of the vaulting conoid with lightweight rubble which would prevent the vault from being distorted.
b) they could make the upper portions of the vault above the conoid more rigid by overlaying it with a layer of concrete almost as thick again as the vault proper.
c) they could use a large and heavy keystone at the apex of the transverse arch and a hanging boss at the intersection of the groin ribs (figs. 22 and 23).
d) they could use ties between the haunches of the transverse arches.

In later Gothic churches, there was more need for tie-rods across the aisles (as at Amiens and Reims) for two reasons:

a) the inward pressure of the aisle vaults against the nave piers.
b) the false bearing of the wall at the back of the triforium passage which, with its superimposed column of long stones set on edge outside the clerestory piers, was part of the stiffening armature of the mature Gothic structural system.

Against these inward pressures there could be no abutment on the nave side comparable to the reverse situation at the level of the high vault where the outward thrust of the vault was received and transmitted by the flying buttresses. In these circumstances the only answer was the use of tie-reinforcement across the aisles.

In Italy, cross-aisle tie-rods commonly appear in churches which have vaulted aisles in association with timber-roofed naves, but here the object of the reinforcement was to prevent distortion of the nave wall by the thrust of the aisle vaults (fig. 21). Reinforcement of this kind also commonly appears in the gynaecia surrounding the dome area in a centrally-planned church as at Fethiye Camii (= H. Mary Pammakaristos) in Istanbul (fig. 41). It appears too in arcades and across naves at levels lower than that of the springing of the high vault (when in wood, this reinforcement was sometimes combined with cross-aisle reinforcement as at S. Maria Assunta, Torcello (fig. 52)).

This type of reinforcement was very common and is found in churches of all periods and areas from Constantinian basilicas to Gothic churches. The object of the reinforcement was to guard against disorders caused by unequal settlement of

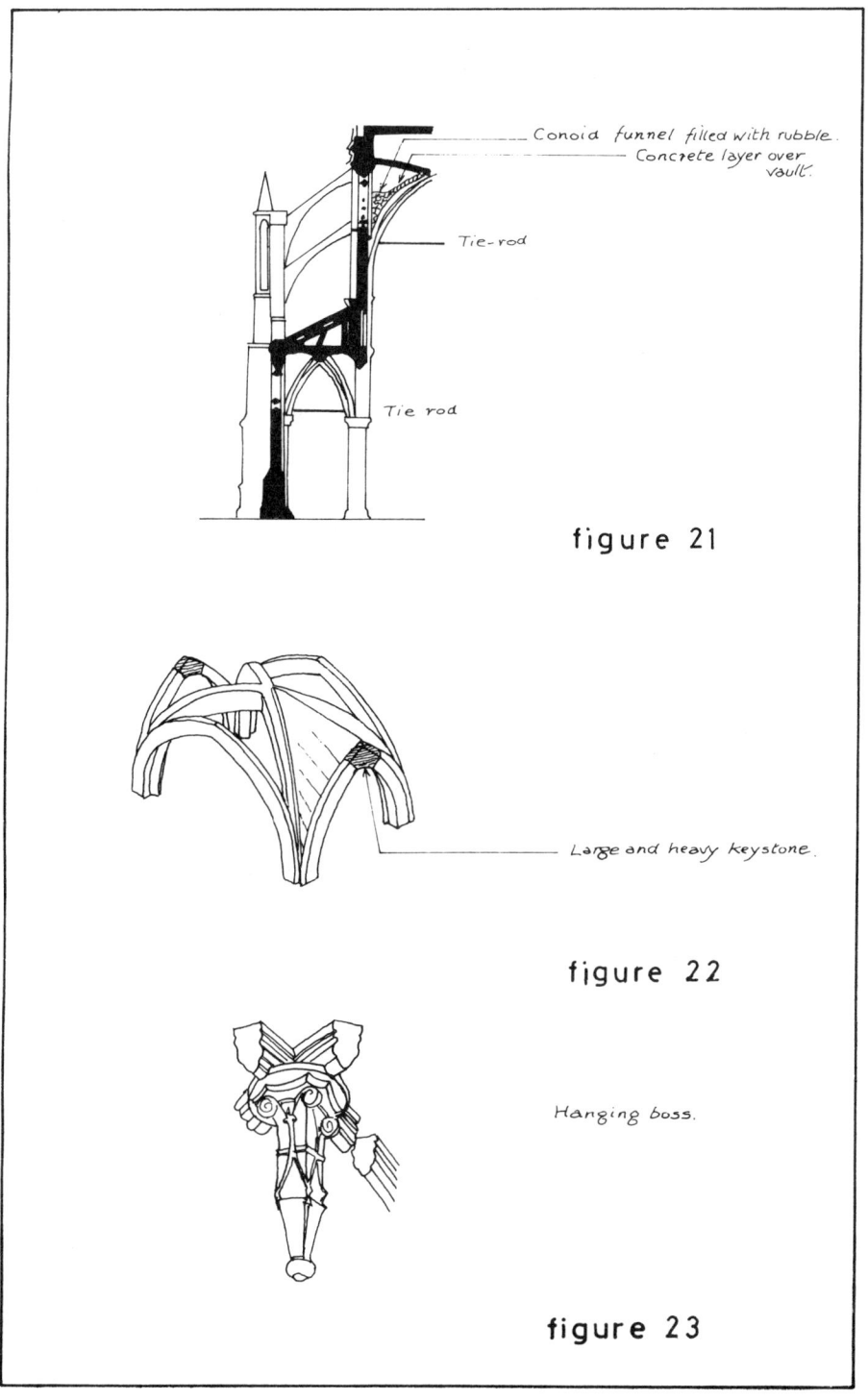

Conoid funnel filled with rubble.

Concrete layer over vault.

Tie-rod

Tie rod

figure 21

Large and heavy keystone.

figure 22

Hanging boss.

figure 23

one part of the building that could threaten to disrupt the whole. The solution to the problem was to ensure that the supports of the buildings—the piers and columns—remained rigid and at a fixed distance from each other and from the exterior walls of the building. This problem was first faced by the Byzantine builders of cisterns in Constantinople with regard to earthquake protection.

A little thought makes it clear that either prop or tie-beams could be used, but in the case of prop-beams it would be necessary to buttress the exterior or far side of an external wall or the final column or pier in an arcade where the end of the final beam is fixed. In practice, the use of tie-beams rather than prop-beams was practically universal in such situations.

The method of retaining prop and tie-beams differed. A prop-beam could be simply wedged into position between the pieces of masonry that are to be kept apart, while a tie-beam has to be fixed into the masonry at each end.

Wooden reinforcement was usually held in position by the grip of the masonry on the ends of the beams. In the case of tie-beams, the weight of the masonry or the bricks surrounding the beam end would compress the mortar between the courses as soon as they were charged by other material above them and this would be enough to hold the beam securely in position in the same fashion as a smoker holds a pipe between his teeth. Tie-beams were sometimes secured into the masonry by metal spikes as at Amiens (fig. 66) or by wooden cross-pieces as at Soissons (fig. 65). Some prop-beams were placed on corbels as at Aquileia Cathedral (pl.10) or held in position by a metal spike as Spilimbergo Cathedral where the abacus was made wide enough for this purpose (fig. 56).

Metal reinforcement could not be held solely by the weight of the masonry, for the compressive factor would powder the stone or brick against the solid metal. The usual method was to pour molten lead into a cavity surrounding the end of the tie. Two other methods were used at Vézelay (fig. 4). Towards the eastern end of the nave, the ends of metal hooks which held the tie-rods were wrapped around intra-mural timbers. At the other end of the nave, the ends of the metal hooks were bent upwards into the masonry (Viollet-le-Duc). Sometimes the rod was pierced right through the wall or passed through a window and was secured by a bar flush against the external face of the wall. This can be seen at the abbey of Vezzolano di Albugnano, Italy (fig. 53 and pl. 29) and at Trogir (pl.12).

The long history of the use of reinforcement in masonry and brick building in the Near East and eastern Mediterranean has already been described. All the buildings cited were buildings with solid walls, with roofs depending, in the main, on continuous walls for their support. It was different in Greece where the use of columns as structural features became more and more common. Such buildings inevitably were more prone to earthquake damage. Usually they were entirely built of masonry with the tops of the columns tied together with a stone architrave. This was the weakest part and presented the least resistance to earthquake shock. Efforts were made to bind together the short lengths of stone comprising the architrave with iron clamps and dove-tail joints. Metal and stone are an unfortunate combination, although it was one that persisted up to and right through the medieval period. The metal corrodes, swells and splits the stone, thus bringing about the very thing it is supposed to prevent. Stone joints will crack apart during seismic disturbances and so provide very little longitudinal resistance to the buckling of walls that occurs at such times.

The only material that was available in sufficient lengths and that could be spliced together sufficiently firmly to provide an elastic architrave was timber. Wooden architraves in combination with stone columns were used in later Greek and Roman times and well on into the medieval period.

> 'The Greek architects set their columns close together, or, in technical language, employed the pyknostyle arrangement; the Romans, choosing to place wider intervals between their columns, were obliged to find some way of distributing the heavy bearing which resulted from the araeostyle construction. Accordingly, they introduced relieving arches, which were at first not open, but hidden in the wall above the architrave. The next step was to show the relieving arches boldly and to substitute a wooden for a stone architrave; and it is easy to see how the widening of the relieving arches would finally do away with the necessity for the architrave altogether'.
>
> (Butler, A.J., 18)

In the quotation above, Butler sets out a logical sequence of development that culminates in the arched arcade set on piers or columns. Butler makes out a case for the validity of this sequence in Egypt; it does not seem to be true elsewhere. But it is worth quoting, not only because a number of the following examples are from Egypt, but also because it is the best exposition of a hypothesis that deals with this particular development in the architecture of the Middle East. Its weakness lies in the fact that not all Egyptian buildings of this period fit into this sequence in their correct chronological order. This is the common fault with most hypotheses of sequential development and can be attributed to the mixture of conservatism and enterprise that characterizes different communities within a given region. In some cases an older tradition is stubbornly adhered to; in others, steps in the sequence are leap-frogged by builders anxious to emulate the latest architectural ideas.

Examples of wooden architraves survive mainly in Egypt, but the best-known is the Church of the Nativity in Bethlehem (530–65) dated by a reference by Eutychius of Alexandria to a rebuilding of Justinian. William Harvey (pp. 9–11) gives a description of wooden beams above the columns of the nave, looking like architraves but bearing no weight. Three beams side by side surround a block of stone which carries the weight of the segmental relieving arches supporting the clerestory walls. Krautheimer makes the point that the church was built by local workmen using techniques traditional to the highlands of the Near East where, as we have already seen, the use of timber reinforcement combined with masonry was an age-old system of building. This building above all others shows the continuity in an area where architecture had been developing longer than anywhere else. The building shows too how these traditions had been modified to suit the demands of particular architectural problems. Here, the weight of the superimposed masonry would have been too much for wooden architraves, so the weight was directed, by means of relieving arches, to the columns, and the wooden beams provided longitudinal stability.

Much less advanced use of timber and columns can be seen in Egypt where conservative tendencies were to be fossilized in most Christian structures by the 'iron curtain' of Moslem occupation which cut off most of the Christian contacts with 'free' Christendom after the mid seventh century. A few of the Egyptian churches had been built by architects who had contacts with architectural practices elsewhere and who took advantage of this knowledge. But others remained conservative even when faced with examples of superior techniques. Such a one was the builder of the monastery of Al Aḥmar (the Red Monastery) at Assiout (Krautheimer; Pococke). Two miles away across the desert was the White Monastery (Deir-el-Abiad) (Gayet) which used beams that only partially supported the weight of the nave walls but, although the Red Monastery was built on the same plan, the older tradition of resting the nave walls directly upon the wooden architrave was used. The church dates from the sixth century, some two centuries later than the building of the White Monastery.

Other examples of conservatism persisted in Egypt. In the seventh to eighth centuries, the church of Ana Shanûdah in Deir-'s-Sifan, between Cairo and Old Cairo, was built with a continuous wooden architrave supporting the nave walls (Butler, A.J.). The architrave performs a decorative function as well as a structural one, for it is carved and coloured. Later still, in the eighth to ninth centuries, the church of Sitt Marian was built (Butler, A.J.). It stands with another church inside the enclosure of the monastery of Deir-es-Suriani (Monastery of the Syrians) in the Wady-el-Natrun in Lower Egypt. Here, too, the wooden architraves carry the full weight of the nave walls. An interesting feature of this church is the use of dosserets (a Byzantine characteristic) above the capitals of the pillars supporting the architraves. This is an unusual arrangement in a Coptic church and it is all the more unusual to find it combined with the conservative method of construction described above. Also in the Wady-el-Natrun is the church of Deir Amba Bishoi (Krautheimer) built during the ninth century. The nave walls rest upon wooden architraves similar to those of Ana Shanûdah. The last of these examples shows a move towards the refinement already introduced in the Church of the Nativity seven centuries before. It is the church of Abu Sargeh (St. Sergius) in Old Cairo (Krautheimer), a colonnaded basilica whose columns are joined by a continuous wooden architrave which rests on the abaci with short, flat pieces of timber intervening to distribute the bearing. The weight of the upper nave wall which rests on the architrave is partially relieved by arched openings of pointed Arab form. The architrave was originally painted. This twelfth-century church is the end of a line of descent whose distinguishing feature is that the timber is integrated with the masonry or brickwork so completely as to form one of the courses (the lowest) of the wall of the nave arcade.

We shall see later how the divorce of the timber reinforcement from the masonry or brickwork leaves the beams free to perform the function of resisting deformations without complicating their function by expecting them to bear superimposed weight. This came about when the weight of upper walls and roofs was directed almost completely onto supporting columns and piers, thus obviating the need for intermediate support. It then became vital to join the supports together so that during a period of disturbance they remained at regular intervals from each other and that any vibrations affecting the building were transmitted throughout the whole structure which could then vibrate as a unit. It was this device that welded a building into a homogeneous entity which was the achievement of the architects of the eastern Roman Empire in its earlier days, but it was not wholly their own invention—the germ of the idea had long been common in different building traditions of the Near East.

Seismic disturbances often have their origin deep inside the earth's crust and the waves sent out are basically of three types, of which two travel through the interior of the earth and the third, which concerns us, passes along the surface. The surface waves are analogous to waves in water and produce an up-and-down motion of the upper strata like waves on the surface of the sea. They die out fairly rapidly with distance from the epicentre with the result that areas close to such centres can be badly affected while others not far off will hardly be aware of the tremor. Certain parts of the earth's crust are particularly prone to earthquakes. Such an area is the eastern and north-eastern part of the Mediterranean basin. It extends as far westward as northern Italy. Places within this area are regularly affected by earth tremors. Outside this area disturbances are infrequent.

In these circumstances, it is natural that architects within the earthquake zone were very much aware of the dangers to which their buildings were liable. It may well be that the tendency towards massive building in the Near East is a reaction to these dangers for, although great wall thickness can be no complete protection, it does safeguard, to some extent, the stability of a building on the fringes of an earthquake disturbance. The damage that does occur can be fairly easily repaired for the basic structure remains. The part of the building that suffers most is the roof for, when the upper parts of the walls sway apart during the disturbance, the roof supports are separated from the wall-plates. The Romans partly solved the problem by constructing monoliths, in which the roofs and the upper parts of the walls were homogeneous concrete units. These buildings were reasonably earthquake-proof, but there is no evidence that the Romans had this in mind when designing them. In Italy, where these buildings originated, earthquake danger never loomed very largely except in parts of the north-east, and this advantage was probably an unlooked-for by-product of the design. In the East, in an earlier period, certain builders approached this particular problem by the development of the stone vault which, by virtue of its weight, sat more firmly on its walls and was better integrated with them than a timber roof would be. This gave a measure of protection against earthquake damage, but was by no means the full answer.

There is, of course, no absolute protection against earthquakes. Modern experience has shown that the best general principle is to make a rigid box strengthened by cross-walls tied in. It is wise to have continuous reinforcement at floor level and under the wall-plate. In bad areas the same should be used at lintel level. Walls should be reinforced and tied to bands and the material used should be concrete reinforced with girders (*Overseas Building Notes*). The nearest approach to this in ancient times is contained in the prehistoric examples quoted in the first part of this article but, of course, the same system could not be used in buildings supported by columns or piers where the walls are not continuous and the rigid-box principle could not be applied within non-existent walls.

The only solution was to tie the tops of the supports together as rigidly as possible with tie or prop-beams or to use iron ties. Wooden beams have the advantage over iron, for they resist both forces acting towards each other (compressive forces) and those acting in diametrically opposite directions. Iron ties are of no use for combating compressive forces and have the disadvantages described in earlier pages. The further advantage of beams is that they can be firmly fixed in the walls.

Despite their disadvantages, tie-rods were extensively used, particularly in the twelfth century in northern Italy. Undoubtedly aesthetic considerations dictated their use. Wooden beams are very often unsightly for, if they are used extensively, particularly across the body of a building, a perspective view from the end of the nave, for example, gives an exaggeratedly bulky impression of their appearance. Iron ties are far less obtrusive.

In order to make wooden reinforcement more presentable, the beams were often carved, which reduced their strength, or were painted, and this was more characteristic of eastern examples, especially the greater churches of Constantinople and its environs. Most of the colouring has faded almost to nothing, but some idea of the original appearance of such beams can be gained in a few of the thirteenth- and fourteenth-century churches of northern Germany whose links with the East via Venice will be traced later. It is a curious fact that these beams when painted were thought to be objects attractive enough to be imitated; there are examples of tie-rods being encased in decorated boxes that look like beams. In H. Sophia, Constantinople and in the church of SS. Giovanni e Paolo, Venice, the tie-rods have been treated in this way and the effect in the latter church is certainly decorative, although to some eyes rather overpowering.

There are examples of masonry beams being used to tie parts of the building together in exactly the same way as timber. They can be seen in the cathedral of Avila in Spain (Harvey, J., 1957) where the apse (begun in 1160) projects beyond the city wall to form a fortified bastion. The projecting wall is tied to the columns of the ambulatory by stone beams. It is a special case for, no doubt, the builders were concerned with constructing this part of the church as massively as possible and such a consideration would explain the use of such unusual material. Other examples occur in a group of churches in Laconia (Krautheimer) which date from the eleventh and twelfth centuries. They all have marble tie-beams in the arches supporting the dome.

Enough has perhaps been said of the background and we can begin to trace the spread of reinforcement of this sort from the earliest known example in a church. This was the church which once stood in Alaşehir in Turkey. The church was dated by Choisy (1883) to between 313 and 350. It had a central dome reinforced with tie-beams both between the piers supporting the dome and across the bowl of the dome where the two intersecting beams were probably jointed together. The beams joining the piers formed a collar that prevented the arches supporting the dome from spreading apart. Higher up at the haunches of the dome the two diagonal beams tied the opposite sides of the dome together. The whole structure of the centre of the church was thus formed into a rigid box which resisted deformations that might have occurred during the period when the church was newly-built and the masonry was settling and also during periods of earthquake disturbance. Little more can be said about the church for it has disappeared since Choisy's time, but presumably it was a domed basilica and a very early example and probably represents one of the first instances of the welding together of Roman building practice and the eastern technique of timber reinforcement. Another building which also contained the same reinforcement and which has also disappeared since Choisy's time was the church at Bursa, also in Turkey.

Early buildings in the new capital of Constantinople were erected in the Roman tradition common at that time to all parts of the civilized world, but there is no doubt that it was already being influenced (as at Alaşehir) by ancient building practices of the Eastern region of the Empire. Later on, such influence blended with the Roman tradition to produce a type of building that, speaking in terms of construction, was decidedly Eastern Roman and later becomes what is recognized as Byzantine architecture. The relative importance of several strands of native influence on the Roman tradition has been fiercely debated for a long time by scholars, but it is clear that some local practical considerations had their effect, whatever the effect of the aesthetic influences was. One of these practical considerations was earthquake protection and the local precaution of timber bonding was taken up and developed. It is in this situation that the genius of the Roman building engineers is seen to its best advantage: a fairly elementary device was adapted in fresh and intelligently-contrived forms. This was not the only technique absorbed from native tradition. The fourth-century buildings have prototypes as early as the second and third centuries in Ephesus, Aspendus, Nicaea (Iznik) and Salonica. The fourth-century buildings used the earlier techniques of mortared masonry levelled off with broad brick bands; solid masonry of alternating brick and stone and brick bands and vaulting of pure brick instead of cast concrete. And so also with the decoration. Krautheimer (79) remarks

' . . . the architectural ornament of the Studios church and, earlier, of the propylaeum of the first H. Sophia (both in Constantinople) is rooted in the building traditions of Asia Minor'.

The earliest examples of the fully Romanized use of two significant architectural features that still remain to us are in the cisterns of Constantinople, built from the mid fourth century onwards. The two features are the pendentive and timber reinforcement. These underground or semi-underground chambers had vaulted roofs supported by many columns. In many cases the columns were reinforced with beams connecting a column with its immediate neighbours. This connection sometimes took the form of single beams, but sometimes up to four were used. The earliest example is the cistern of Aetius (Çukur or Tschukur Bostan) and the drawings (figs. 24 and 26) show the empty beam-holes and that the cistern was built of a mixture of materials salvaged from earlier buildings: the columns vary in length so that some had to be raised on plinths and the capitals are of different types. It may be that the need to weld together this collection of heterogeneous materials was one of the considerations that suggested to the builder the use of reinforcement. Most of the reinforcement ran in the same longitudinal direction along rows of columns. It consisted of single beams set into the springing of the arches which support the vaults, but it was not a complete system for only a few beams ran in the opposite direction. The construction gives the impression of being rather *ad hoc*, but it is interesting insofar as the system of vaulting is an economic and efficient one—a collection of individual domes multiplied to cover a fairly large area. The smallness of the domes makes a large number of supports necessary, but this is compensated for by the great strength given to the structure by this multiplication (Forchheimer and Strzygowski).

A similar cistern is that of Imrahor Camii (Stoudios or Mirachor Dschami) built of the same sort of materials, but with vaulting at a lower pitch (Forchheimer and Strzygowski). It had identical tie-beam reinforcement (fig. 25). Very like this construction is that of Yeri Batan Sarayi (Seray). Here, however, the system of tie-beams is more thorough: in fig. 27, redrawn from Choisy (pl. XIII, i), it can be

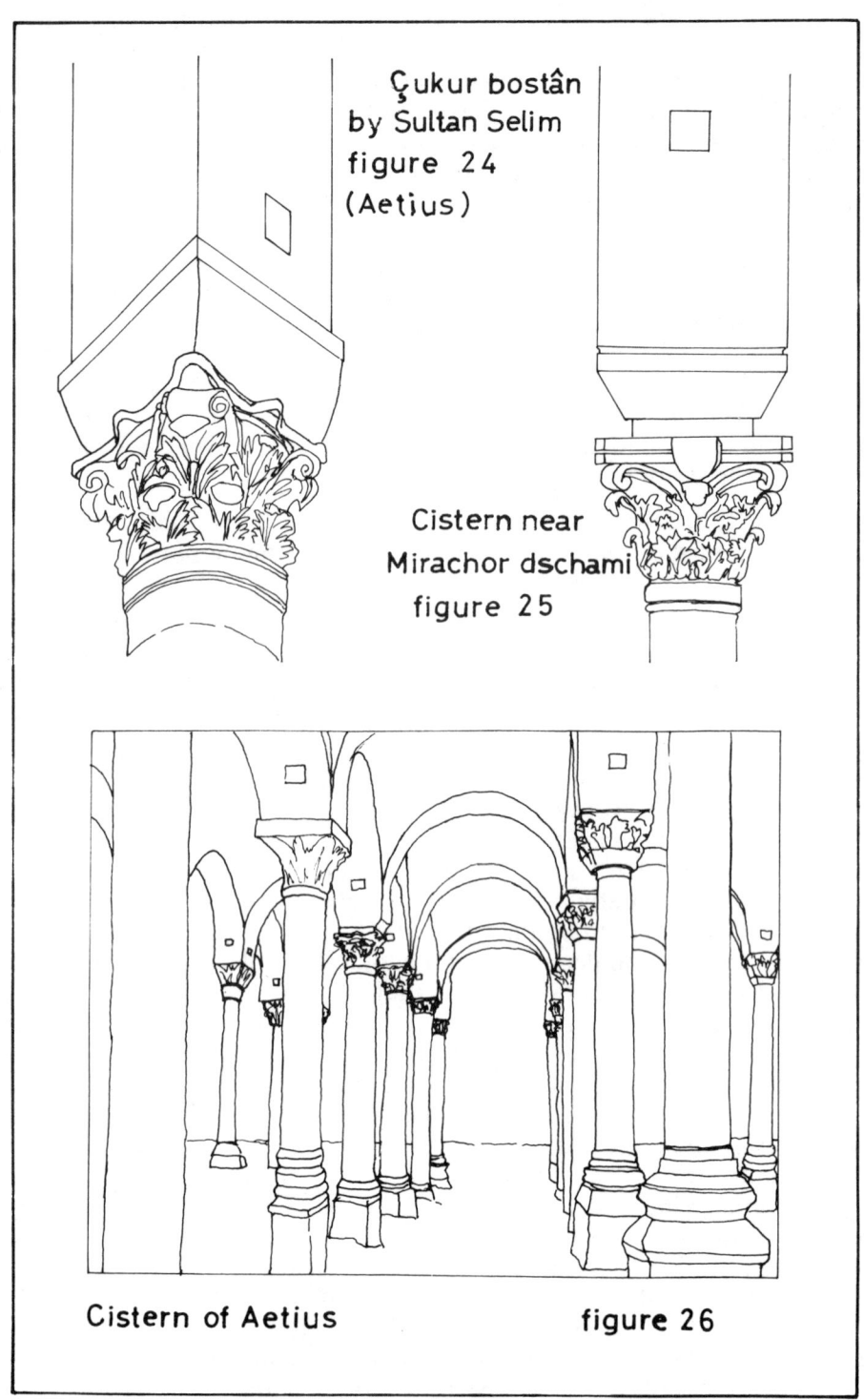

Çukur bostân
by Sultan Selim
figure 24
(Aetius)

Cistern near
Mirachor dschami
figure 25

Cistern of Aetius   figure 26

PLATE 1. Lewes Priory refectory

PLATE 2. Ludgershall Castle. The standing tower from the south.

PLATE 3. Bridgnorth Castle.
Double tier slots.

PLATE 4. Bridgnorth Castle.
The leaning tower.

PLATE 5. Winchester Old Minster. Foundation timber slots.
Copyright Winchester Research Committee

PLATE 6. Winchester Old Minster. Intersection of foundation timber slots.
Copyright Winchester Research Committee

PLATE 7. Lewes Priory reredorter. Chase with rounded top.

PLATE 8. S Maria delle Grazie, Grado. Nave arcade.

PLATE 9. S. Fosca, Torcello. Tie beams in the exterior galleries.

PLATE 10. Aquileia Cathedral. Prop-beams in the nave.

PLATE 11. S. Stefano, Verona. Ends of the tie-beams projecting through the windows.

PLATE 12. Trogir Cathedral. Tie-rod fixings outside the church.

PLATE 13. S. Maria in Organo, Verona. Tie-beams in a nave arcade.

PLATE 14. S. Sepolcro in S. Stefano, Bologna. Restored tie-beams.

PLATE 16. Sv. Chrysagone, Zadar. Tie-beam holes in an aisle.

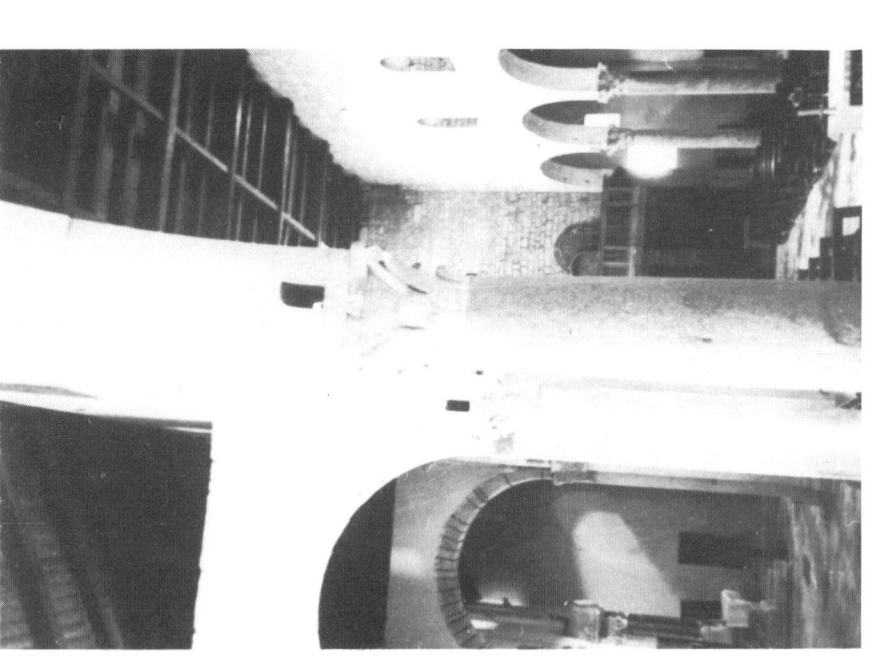

PLATE 15. Sv. Chrysagone, Zadar. Tie-beam holes in a nave arcade.

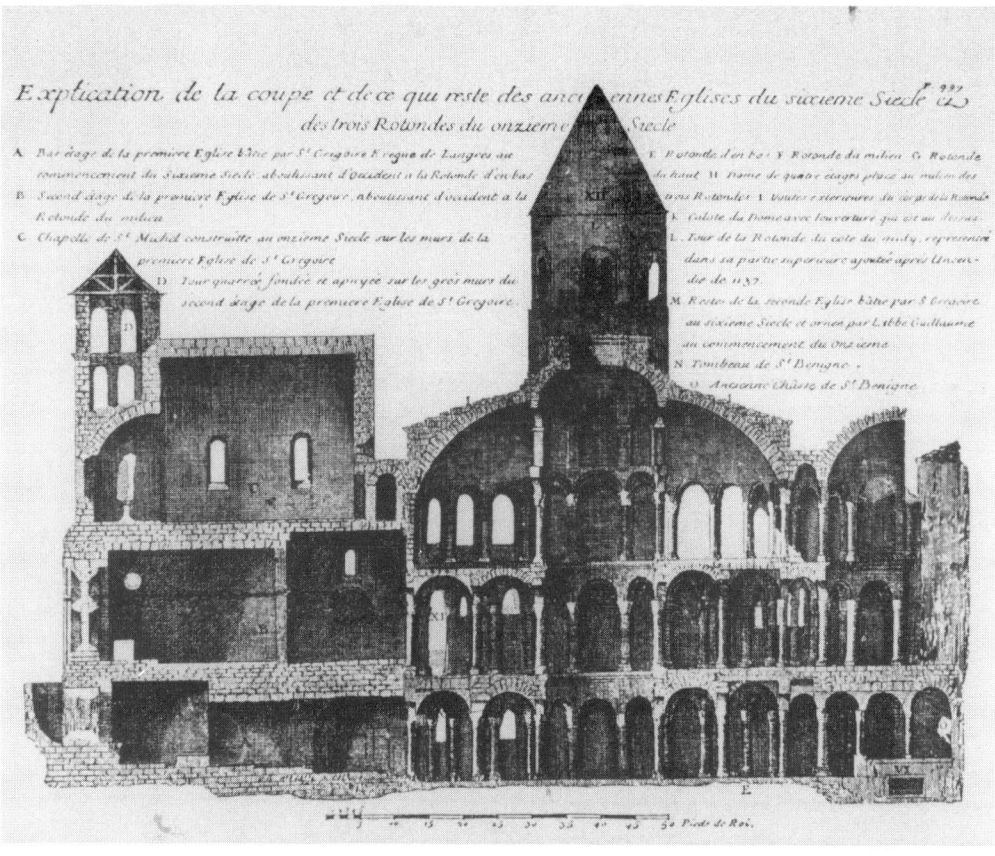

Explication de la coupe et de ce qui reste des anciennes Eglises du sixieme Siecle et des trois Rotondes du onzieme Siecle

A Bas étage de la première Eglise bâtie par St Gregoire Eveque de Langres au
   commencement du Sixieme Siecle aboutissant d'Occident à la Rotonde d'en bas
B Second étage de la première Eglise de St Gregoire, aboutissant d'occident à la
   Rotonde du milieu
C Chapelle de St Michel construitte au onzieme Siecle sur les murs de la
   première Eglise de St Gregoire
   D Tour quarrée fondée et apuyée sur les gros murs du
      second étage de la première Eglise de St Gregoire

E Rotonde d'en bas F Rotonde du milieu G Rotonde
   du haut H Dome de quatre étages placé au milieu des
   trois Rotondes I Voutes exterieures du corps de la Rotonde
K Culotte du Dome avec l'ouverture qui est au dessus
L Tour de la Rotonde du coté du midy, representée
   dans sa partie superieure apostée après l'incen-
   die de 1137.
M Reste de la seconde Eglise bâtie par St Gregoire
   au sixieme Siecle et orné par l'Abbé Guillaume
   au commencement du onzieme
N Tombeau de St Benigne.
   Ancienne Chasse de St Benigne.

PLATE 17. Plancher's section drawing of St. Bénigne.

PLATE 18. St. Barthélemy at Farges (Saône-et-Loire). Tie-beam holes in nave wall.

PLATE 19. St. Barthélemy at Farges (Saône-et-Loire). Tie-beam holes in chancel arch.

PLATE 20. St. Philibert, Tournus (Saône-et-Loire). Tie-beams in the upper narthex.

PLATE 22. St. Charité-sur-Loire (Nièvre). Reinforcement hole in pier of ruined nave (at level of house eaves).

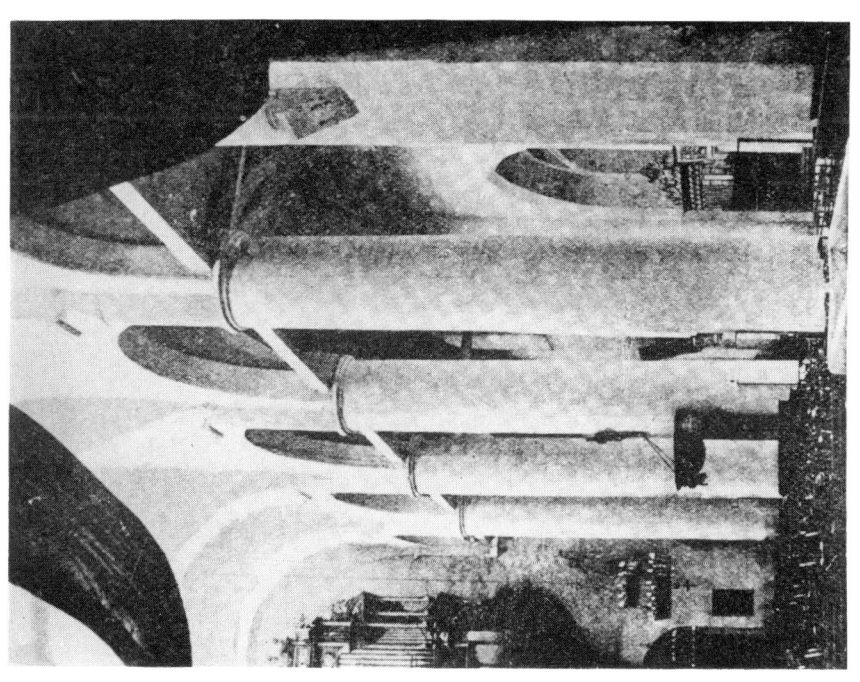

PLATE 21. St. Philibert, Tournus (Saône-et-Loire). Nineteenth-century photograph of tie-beams in a nave arcade.

PLATE 23. St. Philibert, Tournus (Saône-et-Loire). Recent photograph of upper part of nave piers shown in Plate 21 showing blocked tie-beam holes.

PLATE 24. Bussy-le-Grand (Côte-d'Or). Tie-beam holes of earlier reinforcement between later tie-beams.

PLATE 25. Brancion (Saône-et-Loire). Tie-beam holes in nave. (The lower two holes are putlog holes).

PLATE 26. Reims Cathedral. Iron hook for a tie-rod in a nave aisle.

PLATE 27. Noirlac Abbey (Cher). Tie-rod hooks in the nave.

PLATE 28. St.Quirinus at Neuss. Tie-rod across the nave.

PLATE 29. Exterior of abbey of Vezzolano at Albugnano showing tie-rod fixings above aisle roof.

PLATE 30. S. Fosca, Torcello. Double tie-beams in interior.

PLATE 31. Poreč Cathedral. Blocked beam-holes in nave arcade.

PLATE 32. Mausoleum at Concordia. Tie-beam holes.

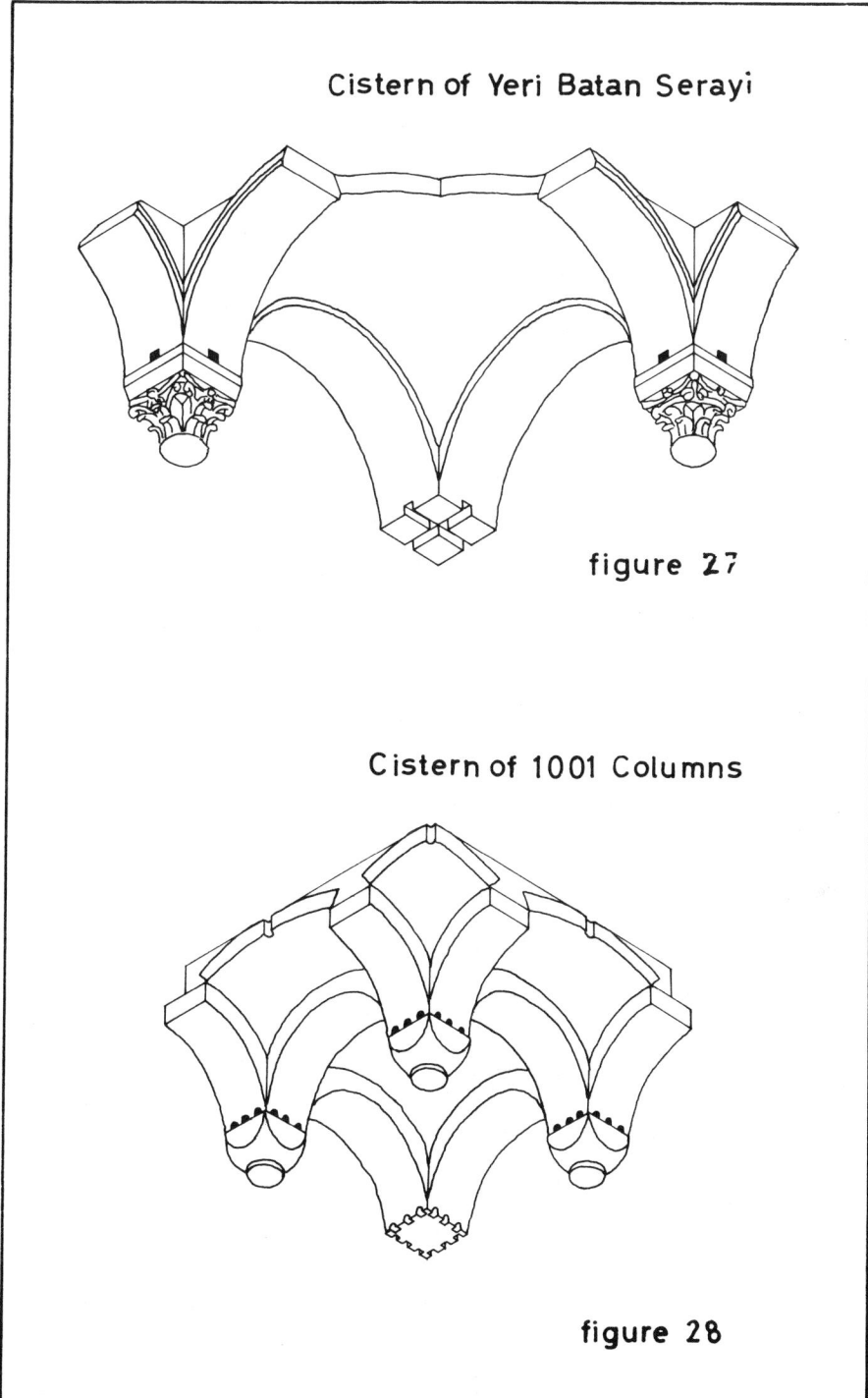

Cistern of Yeri Batan Serayi

figure 27

Cistern of 1001 Columns

figure 28

seen how the four-way tie-beams were jointed together in the interior of the masonry of the springing of the vaulting arches. A complete system of tie-beams like this has the effect of producing a rigid construction that is protected against disturbances from any direction and all vibrations are conducted throughout the whole structure, enabling it to move as a homogeneous unit during earthquake shock.

As the population of Constantinople continued to grow, the number of cisterns proliferated. During the sixth century (528) was built the cistern of Bin-Bir-Derek (1001 columns) described by Diehl as being equal in structural expertise to H. Sophia. This is, no doubt, an exaggeration, but the cistern is undoubtedly a fine piece of engineering. It covers an area of 64 by 56 metres (approximately 70 by 60 yards) and in this space are 212 columns which in some places were joined to each other by triple tiers of tie-beams. Each of the columns carries an early example of capital-impost and supports vaults composed of domes raised on pendentives (fig. 28). The components were all originally produced for this particular construction; there is no evidence of re-used material as in the earlier examples described above. Either the supply of material that could be re-used was running out or a new building programme producing new cisterns from fresh material was concerned with more uniform construction. The multiplication of tie-beam reinforcement perhaps points to a greater concern for earthquake protection. This example was copied elsewhere: Alexandria has a cistern which contained beams of the same pattern (Diehl); another city which used cisterns was Nicaea (Iznik). Fig. 44 shows how the system was arranged: four-way through beams jointed in the heart of each pillar and two other beams jointed at each corner at right angles. This is the cistern of Imbaher and, like the one in Alexandria, cannot be dated, although the beam system inclines one to the view that it might belong to the sixth century, at about the same time as Bin-Bir-Derek.

Of the forty or so cisterns in Constantinople built in the thousand years after 421, fifteen used tie-beam reinforcement. The others, built with less care, had none, and one can make the guarded statement that these appear to be in a worse state of repair. The list below gives the names of those which contained tie-beams, omitting the examples already described. An approximate date is given where possible:

	Cistern	*Approximate date*
a)	Unnamed cistern described by Choisy (1899, 59)	Period between Justinian I and 867
b)	Cistern opposite Ortaçesme	
c)	Cistern near Nuruosmaniye	
d)	Cistern in the Marmara quarter	
e)	Cistern north of Nişanci Cami	Macedonian period 867–1057
f)	Cistern at Köroğlu Sokaği	
g)	Cistern at Otlukçi yokuşu	Comnene period 1057–1185
h)	Cistern near the Fethiye	after 1185
i)	Unidentifiable cistern described by Forchheimer and Strzygowski (p. 87).	

The importance of these cisterns in this paper is that they demonstrate how builders of the period, aware of the earthquake danger and disregarding aesthetic considerations in these utilitarian structures, used what they had found to be the best safeguard and continued to use it throughout the Byzantine period. We shall see also how in church buildings of the area practical considerations over-rode the aesthetic during this same period and, indeed, beyond it, so that timber and iron reinforcement remained a regular feature of the architecture.

Before continuing the story of the development of this technique in churches in the capital of the Eastern Roman Empire, we should look at buildings in other parts of the Empire. It is probably true that the examples with reinforcement that survive in Istanbul are not the earliest that were erected in the city; a good many churches were rebuilt during the reign of Justinian I and it is from this time that we have the first extant example. But, elsewhere in the Mediterranean area, buildings remain from an earlier period. Three of these contain tie-beams. It is probable that two of these were the result of a combination of native and Roman influences like the church at Alaşehir, but the third was probably inspired by Imperial architects from Constantinople. The first of these provincial buildings is Mshatta Palace in Syria (fourth-fifth centuries). The Palace has double tier tie-beam holes in the springing of the arch over the entrance to the principal building (fig. 29). Stylistically the building belongs to the Syrian school (Strzygowski). The second building is the church in the White Monastery (Deir-el-Abaid) at Assiout near Sohag in Middle Egypt. This has already been mentioned in connection with wooden architraves, but it has two features that prove that, unlike several other Egyptian churches, it was built by an architect who was willing to learn from more ambitious examples of church architecture. The plan shows that the chancel of the trilobate cathedral at Hermopolis (Ashmunein) was copied here and evidence remains to show that the lower arcades, supporting the domed vaulting of the nave, were reinforced with tie-beams. The church was built c.440. At Salonica is the church of the Acheiropoeitos (Holy Virgin), a Constantinian basilica dating from around 470. It is one of almost thirty such churches in and around Greece which reflect the style and liturgical demands of the area at that time. The Acheiropoeitos is one of the grandest of these and contains fine mosaics. It was almost certainly built using the most important constructional techniques of the period. One of these was the use of tie-beams and tie-planks (a in fig. 31) in the nave arcades and galleries. It is possible that the church was built by architect from Constantinople; it is like the church of St. John Stoudios (463), and it is certain that the buildings of Salonica were always heavily influenced by architectural fashions of the capital (Hoddinott).

Constantinian basilicas of this type were common in other places than in the Aegean. They are to be found today in other parts of the Roman Empire. Two others that use reinforcement in arcades of the nave are to be found in Rome and in Pula, in present-day Yugoslavia (fig. 50). The Rome example is one of the earliest in the city and dates from between 421 and 432. It is the church of S. Sabina on the Aventine (Krautheimer). Although restored several times since its erection, it still contains the holes for reinforcement in the nave aisles. Interestingly, this reinforcement was of iron, the earliest surviving example of metal reinforcement and the only example of such in a church of standard Constantinian plan. Pula Cathedral is another basilica of the same sort. This church contains tie-beam holes in the nave arcades. Some of these holes contain the remains of

50 *R. P. Wilcox*

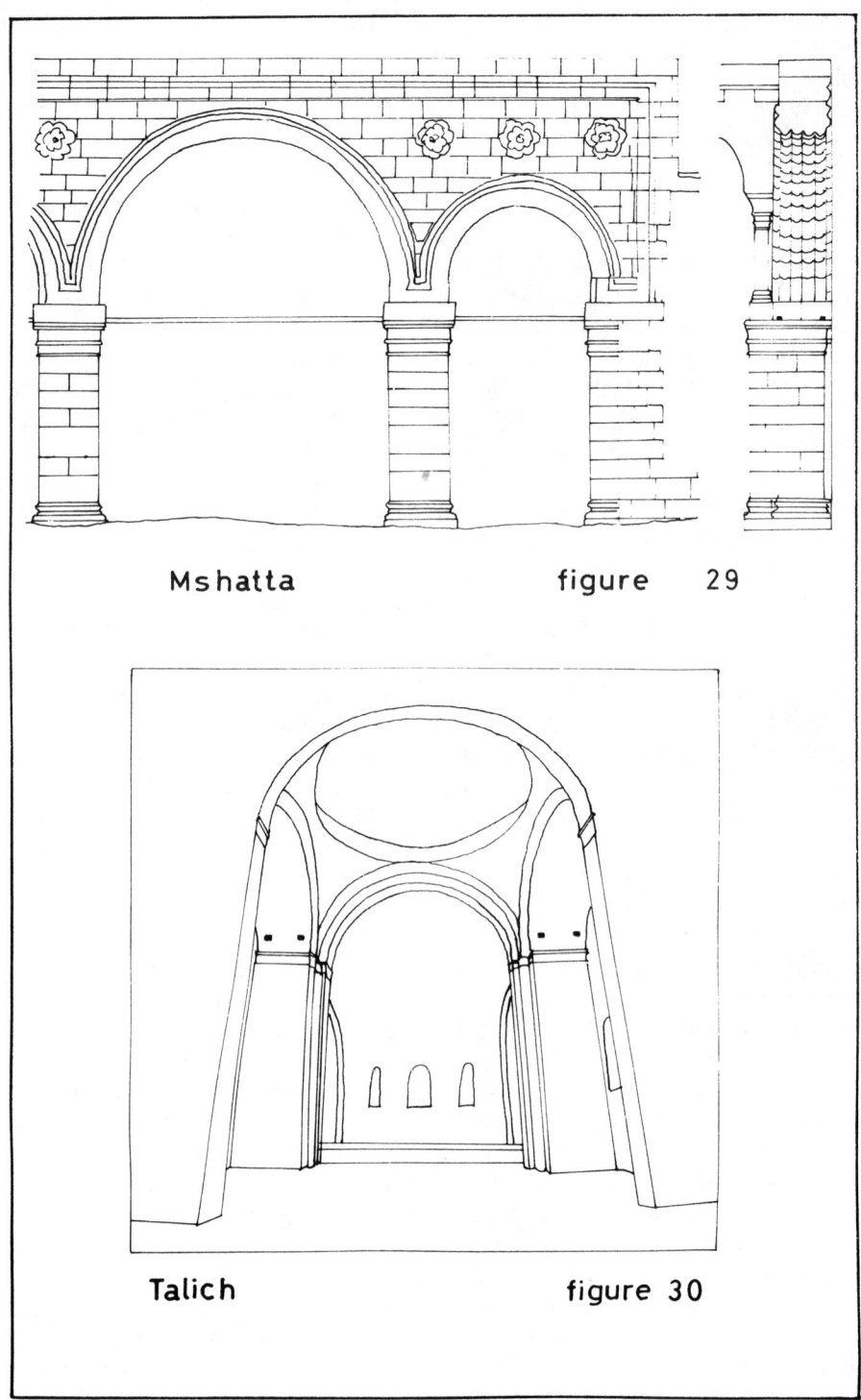

Mshatta   figure   29

Talich   figure 30

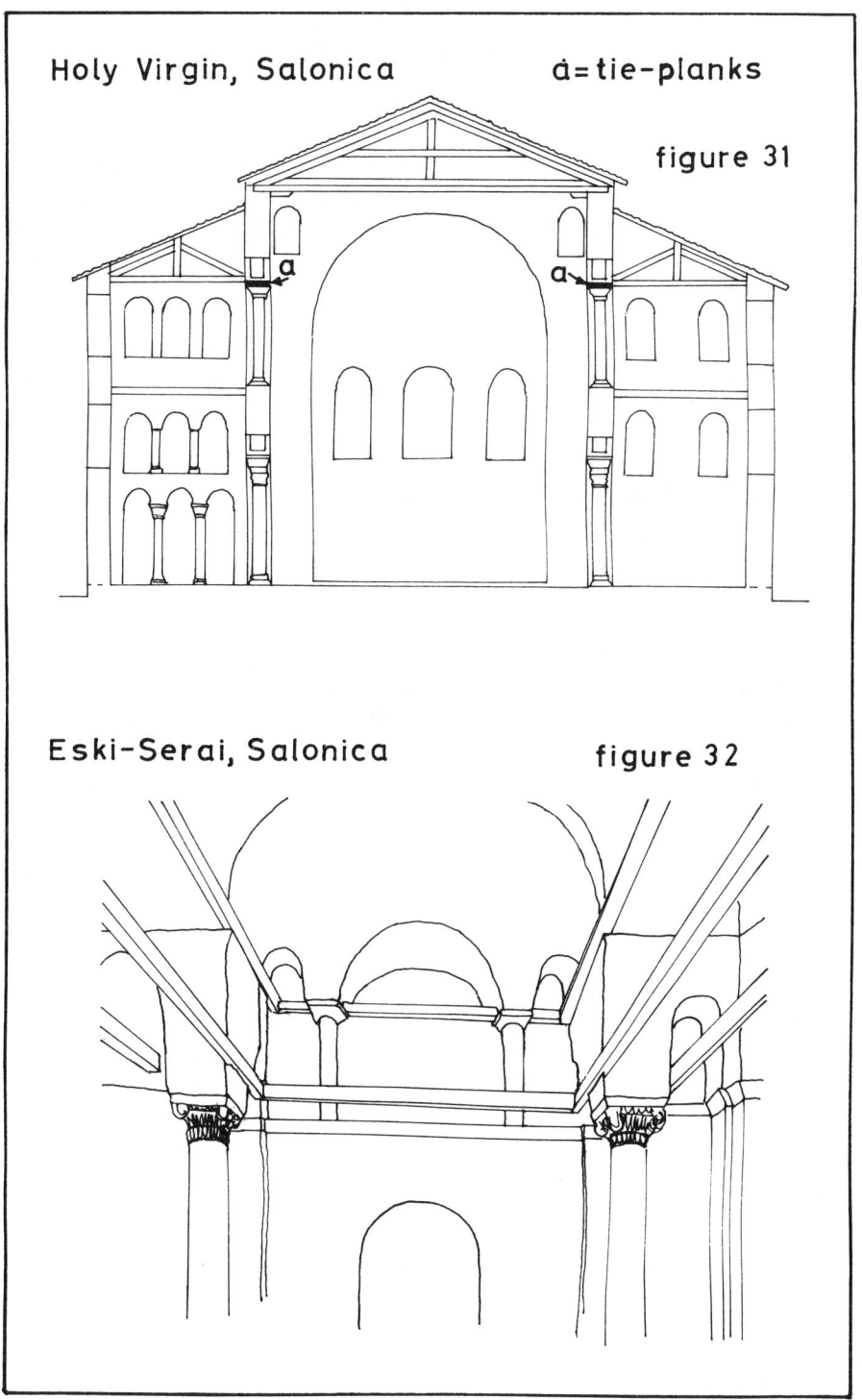

Holy Virgin, Salonica   ȧ=tie-planks

figure 31

Eski-Serai, Salonica   figure 32

sawn-off beams. The dating of this church varies from authority to authority, but its similarities to basilicas in Rome and Milan point fairly strongly to the fifth century, presumably before 480 when the area came under barbarian domination, but it is impossible to get any closer than this. We return to Salonica, to the well-known church of St. Demetrius. It is mentioned here, rather than earlier in this account, because of a difficulty about the date. The cross-transept plan is certainly Constantinian, but the building has suffered from two disastrous fires and it is not certain that the timber reinforcement dates from before the first fire in the early 630s. If one accepts Krautheimer's suggestion of the later fifth century, then the church should be bracketed with the Acheiropoietos, but other authorities state that it was completely rebuilt after the disaster. The church is a basilica built in a style that reflects, like the Acheiropoietos, the proximity of Constantinople. It was restored after the second fire in 1917 and it now contains tie-beams both in the lower and the upper nave arcades. Before this fire, tie-beams existed in the lower nave arcades, but the upper arcades were reinforced with tie-planks like those in the Acheiropoietos (fig. 33).

In Constantinople, developments during the reign of Justinian resulted in a number of churches which involved the adoption and use of several architectural traits that had not been brought together before. These developments quickly spread throughout the eastern Roman Empire and penetrated to the West under the influence of the political expansion and consolidation that was carried through during Justinian's reign. The Eastern Empire regained Italy and North Africa, bringing a revitalising spirit in its wake which arrested the artistic decline associated with the economic and intellectual degradation of the West at that time. The evidence for this revitalisation is one-sided; it is really only in architecture that it can be readily demonstrated. It is for this reason that any evidence for direct Eastern influence is important and part of this evidence is the reinforcement in some of these monuments. Decorative details and church plans could have been transmitted at secondhand, the first by means of works of art small enough to be carried by way of trade, and the second by means of drawings, but constructional techniques like reinforcement could only have been transmitted by builders who had actual experience of eastern church construction. This consideration is the kernel of this thesis.

Four kinds of churches were built during the sixth century. The timber-roofed basilica continued with only slight modifications of the type of the preceding century. Secondly, there were centrally-planned buildings whose shape was either circular or octagonal. H. Sergios and Bacchus (Küçük Ayasofya, 527–36) is an example in Constantinople and S. Vitale (547) another in Italy. These two plans were combined in a third type, the domed basilica. An example in Syria is Qasr Ibn Wardan (561–4). The fourth type used the cross-in-square plan which was introduced during the reign of Justinian, but which did not achieve full development until later.

We have already looked at basilicas and seen how these were sometimes reinforced with tie-beams and tie-rods. An example from the sixth century is Poreč in Yugoslavia (fig. 34 and pl.31); in Constantinople no examples at all date from this time. Poreč (*c.* 550) was probably built by local workmen, but contains capitals and columns provided by Constantinople. This is characteristic of the strong influence on the northern shorelands of the Adriatic exerted by the East

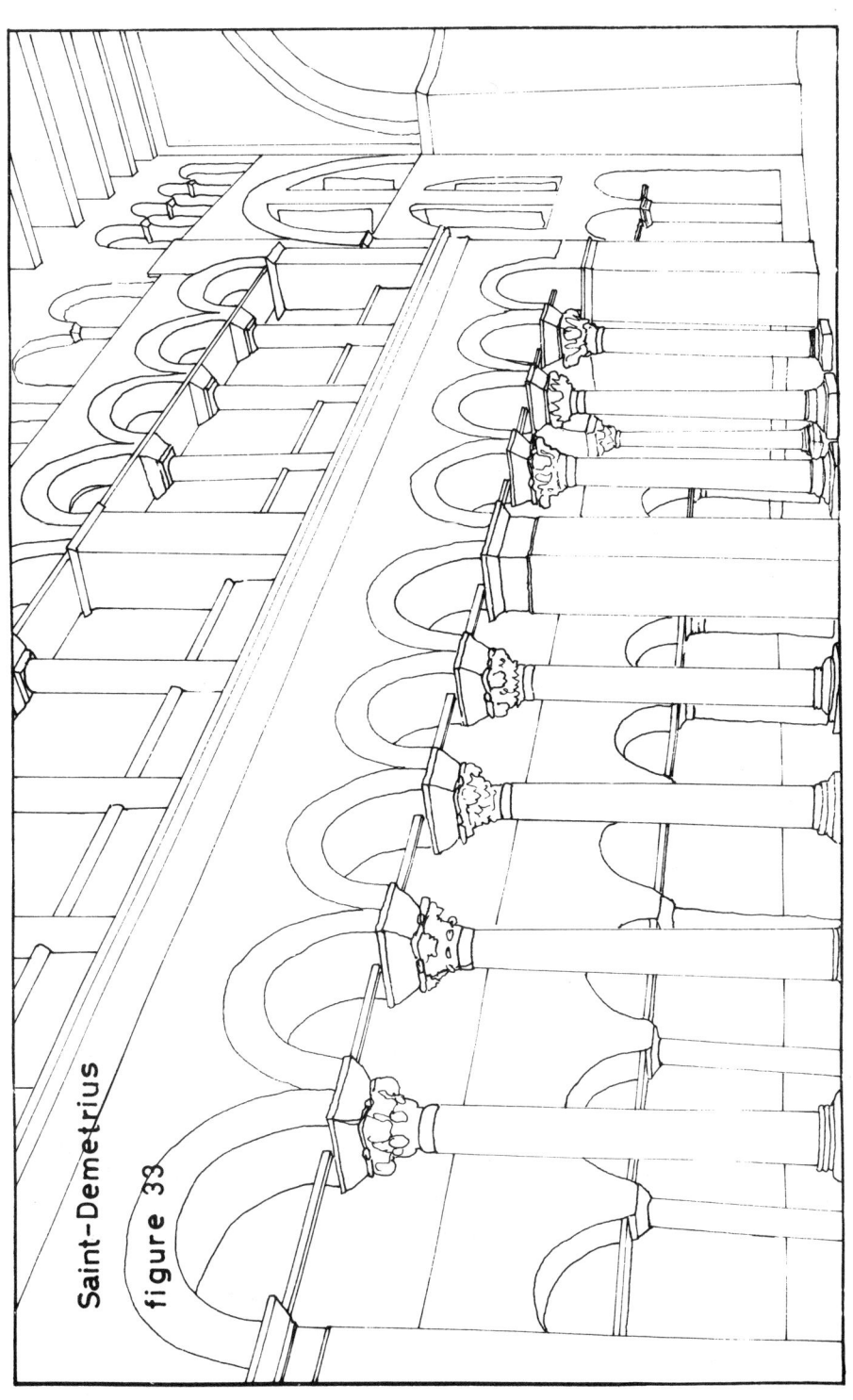

Saint-Demetrius

figure 33

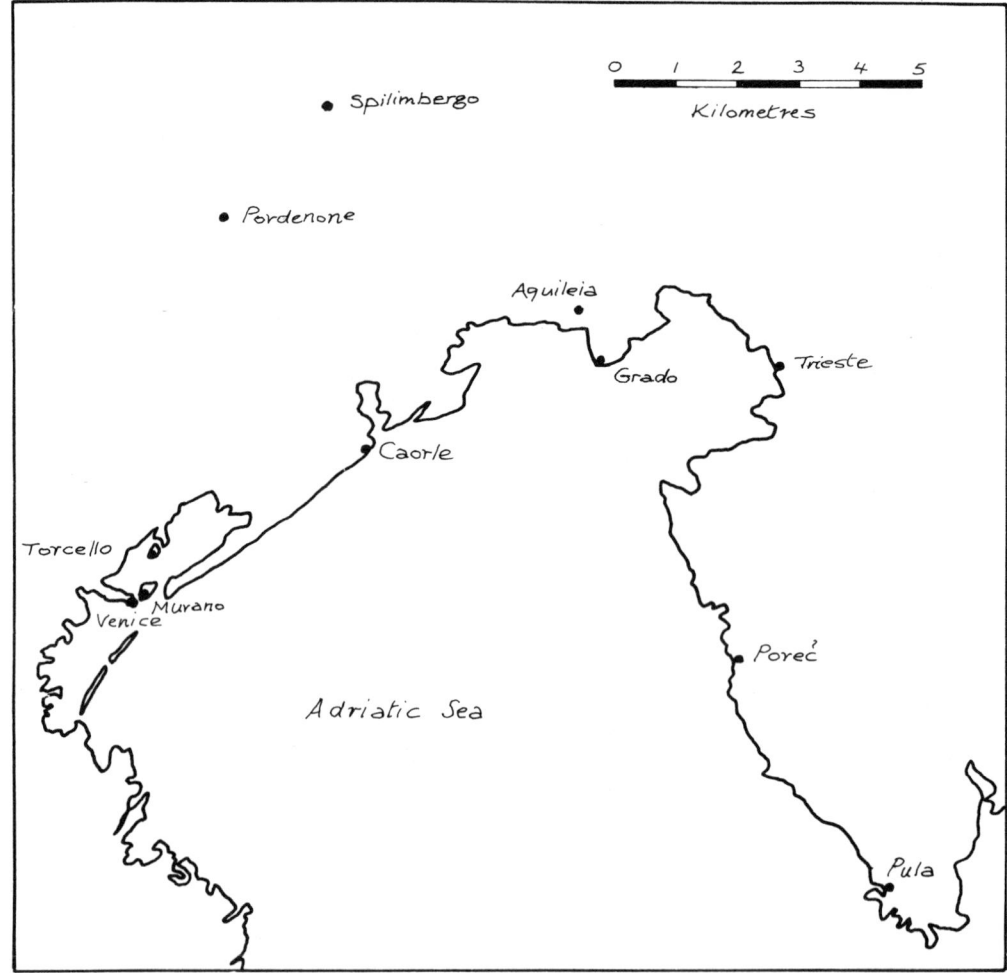

FIG.34. Map of the northern littoral of the Adriatic

which started in the mid fifth century and never entirely died out thereafter. In the church, timber reinforcement was used across the nave. The blocked holes of this reinforcement can still be seen in the walls (pl.31).

Of the second type of church, the centrally-planned building, the early example of H. Sergios and Bacchus (Van Millingen) still stands in Istanbul. Its plan is in the shape of an octagon surrounded by a square wall. On the western side is a narthex and it is in its upper gallery that the first surviving example of iron reinforcement in the East appears. Tie-rods are laid through springers of the arches and join together piers on whose abaci they rest (fig. 35). Originally, the church formed part of a private palace of Justinian and to this fact we owe a notice by Procopius (*De Aed.* I. 4) whose description allows us to date it firmly and to picture it as it originally stood, gleaming externally with marble and internally with gold. It is possible that the other magnificent church mentioned above, S.Vitale in Ravenna, used iron reinforcement, but the evidence is not strong and on the analogy of the other fine churches in the city that do not contain reinforcement, one should do no more than simply mention the possibility.

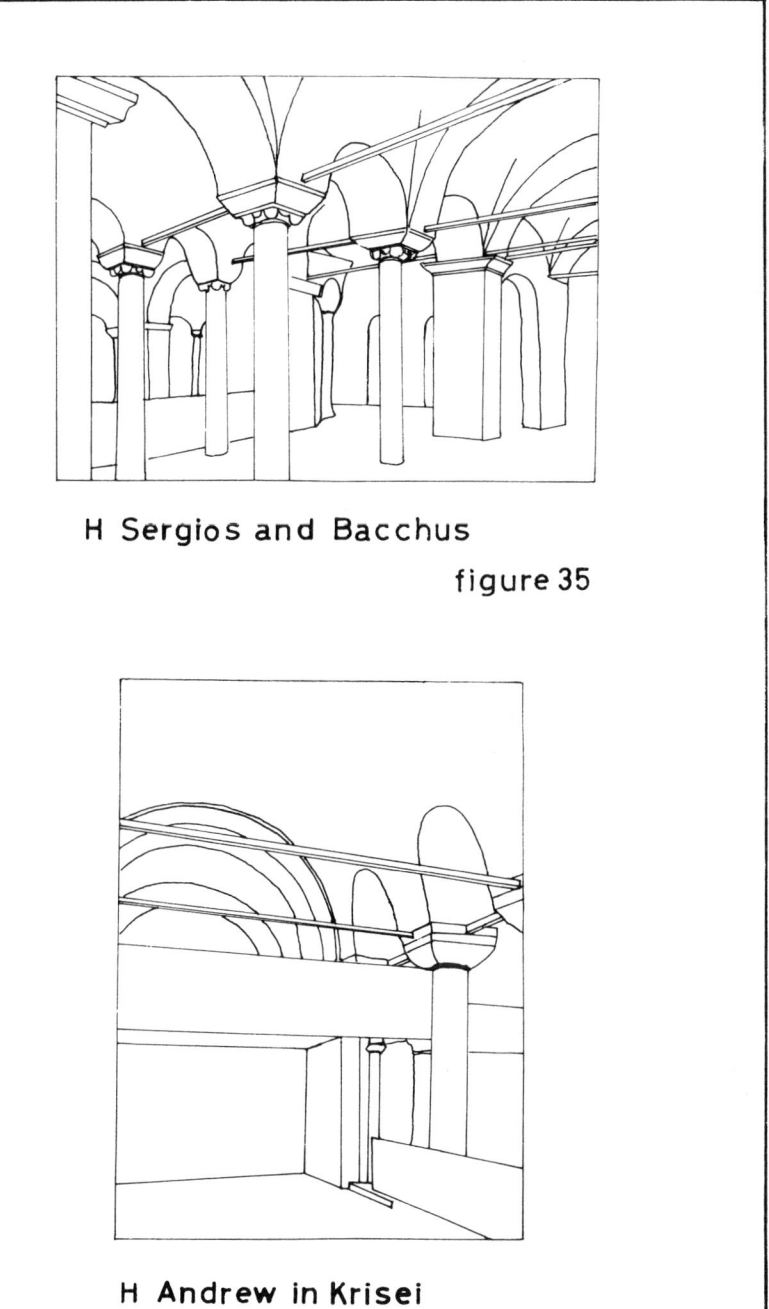

H Sergios and Bacchus

figure 35

H Andrew in Krisei

figure 36

The origin of the domed basilica cannot be determined today. Tentative experiments with the plan were made in Asiatic Turkey at Koja Kilisse in Isauria and in Cilicia at Meriamlik and at the White Monastery in Egypt during the fifth century. It will be remembered that the White Monastery contained tie-beams under the domed vaulting of the nave. As a general rule, the churches in Turkey had domes over the nave while the Egyptian examples had cupolas over the apses. In the church of H. Sophia (Ayasofya) at Constantinople the domed-nave plan was taken up and enlarged (Kautzsch). The dome is the central factor in the whole ensemble. The rest of the building simply surrounds it and supports it. So effective is the support provided by abutting half-domes, semi-circular arches and four enormous piers, that the central dome appears to float on air. There is an immense amount of tie-reinforcement in the church. In the collaterals, tie-beams and tie-rods are extensively used while the galleries are reinforced with tie-rods encased in decorated wooden boxes (figs. 37 and 38, redrawn from Lethaby and Swainson). The lower part of the narthex also contains tie-beams. How important this reinforcement was can be gauged from the effect of an earthquake on the church in 538. Part of the dome fell, but the building still remained standing. Since the reconstruction necessitated by this disaster, the church has stood comparatively unscathed despite several other earthquakes. H. Sophia represents the highest achievement in Byzantine architecture and as such was built using the most superior techniques of the day and its influence was immense. It was the subject of exorbitant adulation and, although nothing on the same scale was ever attempted again, its constructional techniques were imitated. The prestige of H. Sophia, therefore, provided a significant boost to the adoption of these techniques and these included both tie-beams and tie-rod reinforcement. H. Sophia was built between 532 and 537.

Another fine church in the capital is H. Eirene or St. Irene, a domed basilica in which the dome is over the chancel and the nave is covered with a barrel vault (Krautheimer). There is a certain amount of dispute about the date of this church. It was certainly rebuilt by Justinian after a fire in 532. In 564 it was again damaged by fire and restored and in 739 it was severely damaged by an earthquake. The point at issue is whether the church was entirely rebuilt after this or merely extensively repaired. On this question the position of the tie-beams is illuminating. The following is a quotation from George (11):

> 'As far as the main walls and vaults are concerned, there is no apparent trace of any system of chainage by wood or iron ties, such as were often built in to secure Byzantine structures against earthquake, but most of the smaller vaults, namely those of the narthex, aisles and western gallery, originally depended to some extent upon wooden ties which have since decayed or have been cut away, with the result that some of these vaults have fallen and have been rebuilt, and the remainder are partially deformed'.

This evidence does suggest that the main vaults were rebuilt c.740 while the remainder of the church was merely restored. For some reason the rebuilt vaults did not contain reinforcement. If this is so, then the parts of the church with evidence of tie-beam reinforcement would date from the Justinian rebuilding of 532 or from 564.

One provincial church that bears evidence of strong influence from the capital is

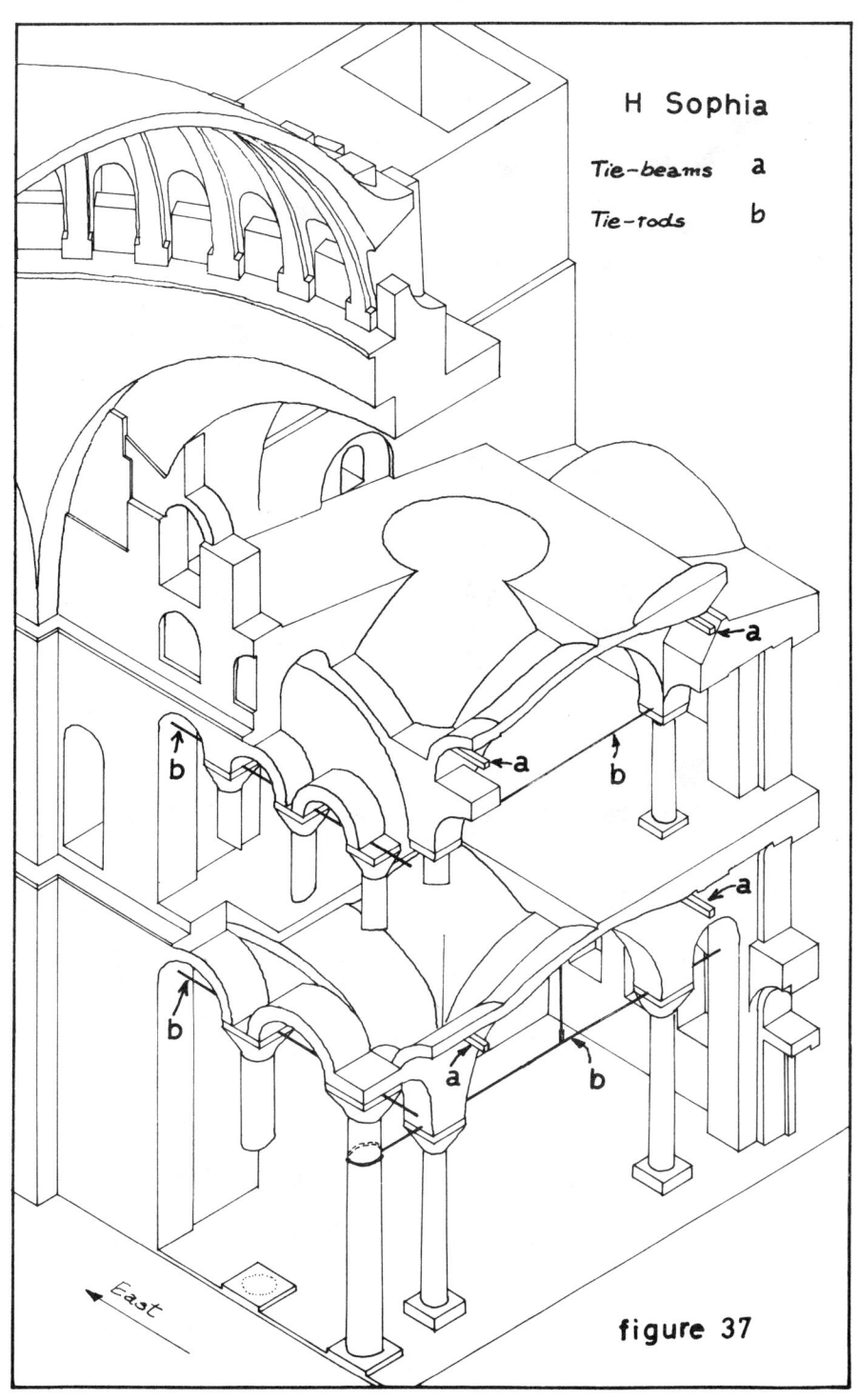

H Sophia

Tie-beams a

Tie-rods b

East

figure 37

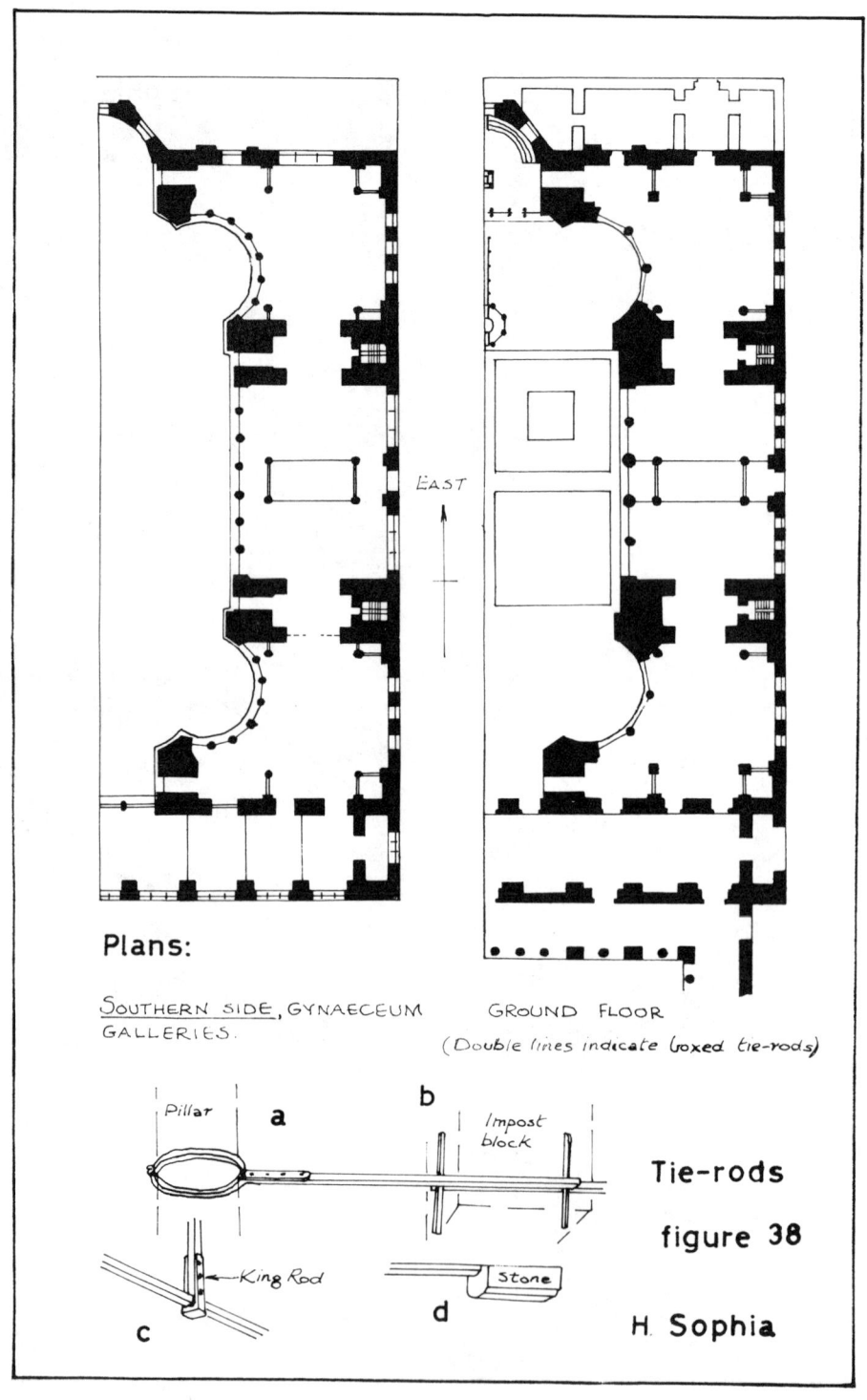

Plans:

Southern side, Gynaeceum
Galleries.

Ground floor

(Double lines indicate boxed tie-rods)

Pillar

a

b

Impost block

King Rod

Stone

c

d

Tie-rods

figure 38

H. Sophia

the church adjoining the palace at Qaṣr Ibn Wardan in Syria. It is a domed basilica with a narthex constructed of brick and local basalt with double tie-beams across the barrel-vault adjoining the central dome. Plan, style, materials and the richness of the decoration recall Imperial buildings at Constantinople and are quite different from the stone churches that had been the rule in Syria in earlier years.

It will be convenient here to trace the influence of the four types of churches in their reinforced forms throughout the Eastern Empire during the rest of the medieval period before attempting the more difficult task of showing something of their influence, in terms of reinforcement, upon the West.

Perhaps the last important outside influences that penetrated Egypt, before the tide of Moslem conquest swept across it and washed away the Eastern Roman army at the battle of Heliopolis in 640, were those associated with the Justinianic period and the years immediately afterwards. The churches were all types of basilicas with stone roofs, sometimes barrel-vaulted, sometimes vaulted in domes. A typical example of the former was the church of Deir-al-Azâm near Siout, built *c*. 600 and completely altered in a rebuilding of the 1870s (de Bock) during which the tie-beams across the choir were removed. Much later, but of the same standard Coptic type is Deir Baramous in the Natrun Valley, dating from the twelfth century (Krautheimer). Here the nave and aisles are barrel-vaulted and supported on piers, the fore-choir resembles a transept with a transverse barrel vault and the chancel is composed of three apses. The choir is reinforced with transverse tie-beams.

Other churches were roofed with domes, like the church in Deir-es-Suriani which stands beside Sitt Marian (already mentioned). This church has domes over the choir and sanctuary and a square, domed bay instead of an apse (Krautheimer). The nave, however, is barrel-vaulted and reinforced with transverse tie-beams (fig. 40). The church was probably built in the ninth century, but the vaulting was not constructed until *c*. 1200, and it is not possible to say whether the tie-beams belong to the earlier or later vault. A later and undateable addition to Al Aḥmar (the Red Monastery) used timber reinforcement below the dome (fig. 39).

Here we leave Egypt, with the stylistic development of churches arrested at a point somewhere in the sixth century. We can now turn to churches elsewhere in the Eastern Roman Empire, starting with the churches of Constantinople which, in most cases, provided the inspiration for developments in other places. The basilica no longer plays any part in the ecclesiastical architecture of the capital, so we shall be concerned with two of the other three types. A variation on the domed basilica is the cross-domed church. Two such churches contain iron tie-rods. Gül Camii (possibly H. Theodosia) has metal reinforcements in combination with barrel vaults. Tie-rods connect together the arcade piers and are set across the nave and below the dome. Evidence for dating is almost non-existent and it could have been built at any time between the mid ninth century and 1200. Another church about which there is some dispute over the chronology is St. Andrew in Krisei (Koša Mustafa Pasha Cami). It is often dated to the sixth century (Van Millingen and Hamilton) but another authority (Krautheimer) now considers it to be of thirteenth-century date. This church is lavishly provided with metal reinforcement. It can be found in a cloister, between the arcade piers and across the nave, while tie-beams are used in the inner narthex (fig. 36).

Of the cross-in-square churches in Constantinople, one of the earliest remain-

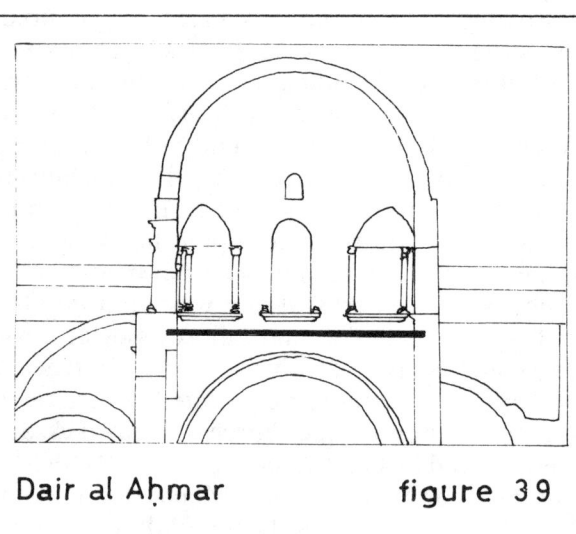

Dair al Aḥmar   figure 39

Al 'Adra, Dair-as-Suriani.   figure 40

ing is St. Saviour Pantepoptes (Eski Imaret Camii). It has iron tie-rods in the narthex supporting a barrel vault and tie-beams below the central dome. The church dates from 1081–1118.

The cross-in-square became the typical plan of later Byzantine architecture. The exterior walls of the church are composed of a square enclosure from the four walls of which vaults reach towards the centre where they terminate in four semi-circular arches. Between these arches, pendentives support a drum which is cylindrical inside and polygonal outside. On this drum is a central dome. The four vaults and, of course, the dome, rise to a higher level than the roofing of the angle compartments in the corners of the exterior wall and can be covered with either barrel, cross or domical vaults. A central apse and two side apses are at the east end and often the smaller apses do not project beyond the eastern exterior wall. Such a design produces a building which is in all essentials a compact unity, in which stability seems to be completely assured in normal conditions. This is probably so, but it has already been pointed out that bricks and mortar have little cohesion in times of earthquake tremor and we find reinforcement used in a great many of the finest churches of this type. The reinforcement was used mainly across the barrel vaults of the arms of the cruciform part of the church.

St. Saviour Pantocrator is one of three churches in the Zeirek Čami and was the funeral church of the Comnene dynasty. As such it was a very important church and was built with the necessary magnificence between 1118 and 1143 (Van Millingen). Unlike a great many churches of this type, it was partly groin-vaulted and used iron tie-rods (fig. 42). Built on a site outside the main city, St. Saviour in the Chora (the Country) or Kariye Camii was very extensively reinforced with wooden tie-beams. There are beams across the arches below the dome and in the narthex. The church has been variously dated by several authorities but the weight of opinion seems to point to somewhere at the turn of the eleventh and twelfth centuries. A last and much later church is St. Mary Pammakaristos (Fethiye Camii) where tie-rods are used in the narthex and in the paracclesion (fig. 41) This church dates from between 1303 and 1321 (Krautheimer). An example of the use of reinforcement in a grand civil building is in the Palace of the Tekfur Sarayi (Constantine Porphyrogenitus). Today, the lower arcade of the north-western facade has been reinforced with modern T-section girders but, in the interior, it is possible to see that tie-beams were originally used in the springing of the arches at right-angles to the facade. The building is dated to the early four-teenth century.

Outside the capital, in other parts of what is now Turkey, there are a number of churches which contain reinforcement and which, apart from the first example, are all buildings which had received inspiration from Constantinople either directly or indirectly. Churches of this region of Asia Minor inherited a tradition of building in large stone blocks and this was the material still used in later periods, although the style and plan reflected the fashions of the capital. The technique of rein-forcement was a device, too, that was indigenous to the area, so, as far as influences are concerned, was neutral, common by now to both the provinces and the capital. An example discovered and described by Ramsey and Bell is the monastery they numbered 43 at Maden Shehir. It is not clear on what plan the church was built but, in the part that was probably barrel-vaulted, tie-beam holes were found. There is practically no dating evidence, but it was suggested that the

H Mary Pammakaristos   figure 41

H Saviour Pantocrator

figure 42

monastery dates from before 700. In the Turkish province of Diyarbakir, in the town of Silvan, formerly Mayyafarikin, is the church of the Virgin, dating from the last part of the sixth century. It was built on a cross-domed plan with a timber roof and tie-beam holes above the capitals amongst the ruins demonstrate the use of reinforcement. It is interesting that a church of this kind, much influenced by Constantinople and perhaps Salonica (for it much resembled H. Sophia in that city) should have been built by the Persian king Chosroes II. Also in Armenia there are the remains of the cathedral at Talich (Thalish), a basilica with a central dome and two side aisles. Below the dome (fig. 30) double tie-beam holes show that wooden reinforcement spanned the blind, deep arches on each side. It is clear here, too, that influences from the western part of the Eastern Empire were important, and particularly those from Constantinople. As Krautheimer remarks (233)

> ' ... the hall churches of Armenia might well be interpreted as a variant of Justinian's cross-type superimposed on the building tradition of the inlands of the Near East'

In south-western Asia Minor, during the ninth century, a church at Dere Agzi was modelled on the church at Qasr-Ibn-Wardan (*c.*564) and, can be described as a compact domed basilica (Krautheimer). It contained tie-beams. Two churches containing timber reinforcement can be dated to about the same time. Both are described by Ramsey and Bell. Canlikilise (Tchangli Klisse) church and Ilanlikilise church were both cross-in-square churches, with a dome over the crossing and barrel-vaults over the rest of the church. Both contained tie-beams; the holes still exist in both churches in the springing of barrel vaults adjoining the crossing, but Canlikilise has double holes while Ilanlikilise has single holes.

The provincial city of the Eastern Empire in which most Byzantine churches of all periods survive is Salonica (Thessalonica). Its basilicas have already been described but there are several centrally-planned churches. The cross-in-square church known as the Theotokos church dates from 1028 and is shown in fig. 45. This transverse section (from Diehl, 1918, pl.LII) shows that the church was reinforced with timber across the nave at the crossing and across the aisles. There are also tie-beams in the nave arcades. The drawing makes clear how the beams must intersect and be jointed together in the masonry above the capitals. A later church on the same plan, H. Pantaleimon, shows the technique in a more elaborate form: the beams are used as at the Theotokos church together with a further tier at a higher level below the dome. This church was probably built during the mid twelfth century. St. Catherine contains further elaboration including beams in the narthex (fig. 46, redrawn from Diehl, pl. LXI). This church dates from the late thirteenth century. Tie-beam reinforcement was used in the chapel of St. Nicholas the Orphan constructed at the turn of the thirteenth and fourteenth centuries. One of the finest churches in Salonica is the Church of the Apostles. Built on the cross-in-square plan, it is bigger than the churches just discussed and is timber-reinforced throughout the interior. On the exterior, the facade incorporates arcades which are also reinforced in the same way. The church dates from the early part of the fourteenth century. The last example in Thessalonica exhibits a development in the use of tie-beams which is not common. This is in the interior of H. Elias (Eski-Serai), built between 1360 and 1380 (Jackson). Tie-beams are

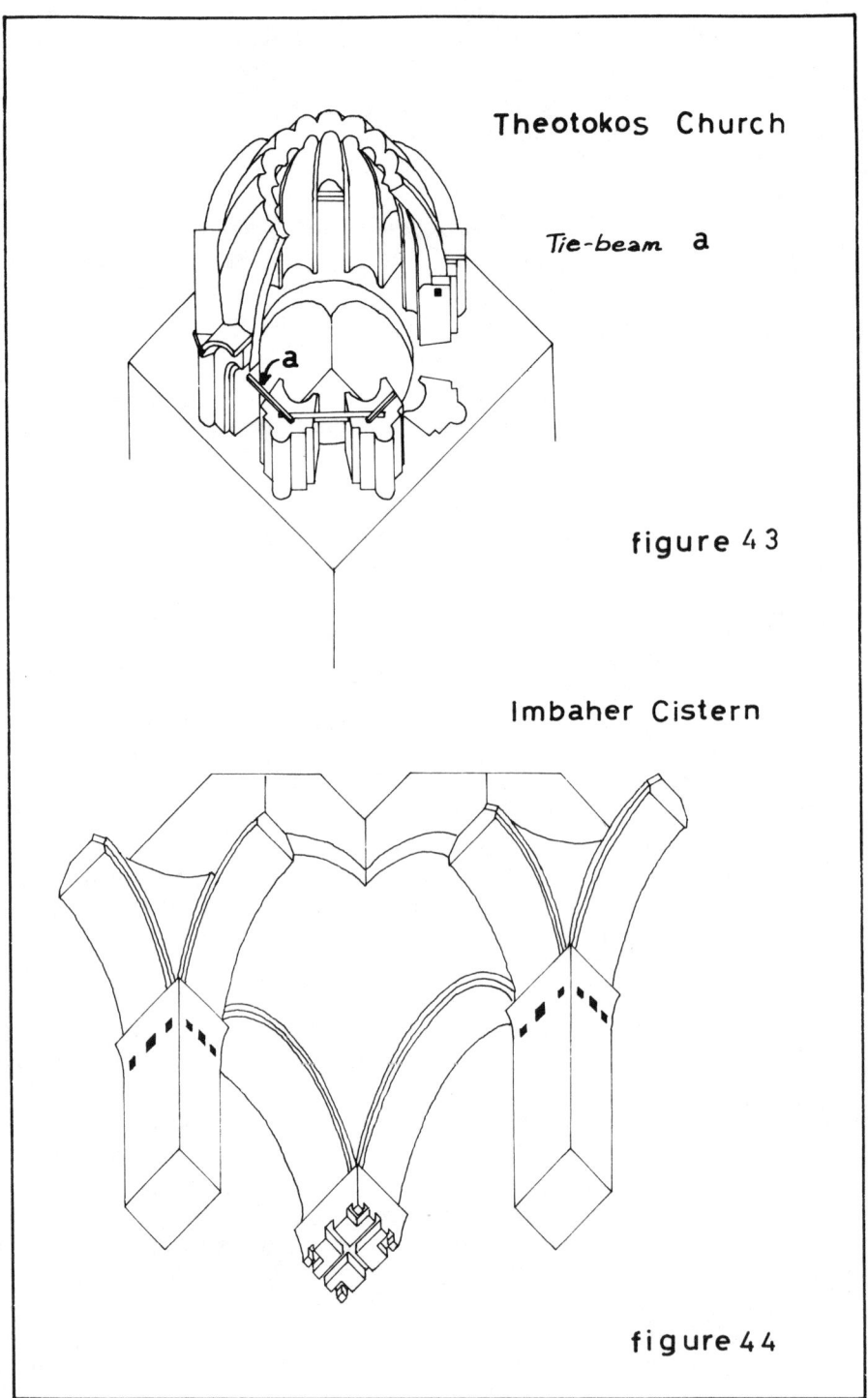

Theotokos Church

Tie-beam a

figure 43

Imbaher Cistern

figure 44

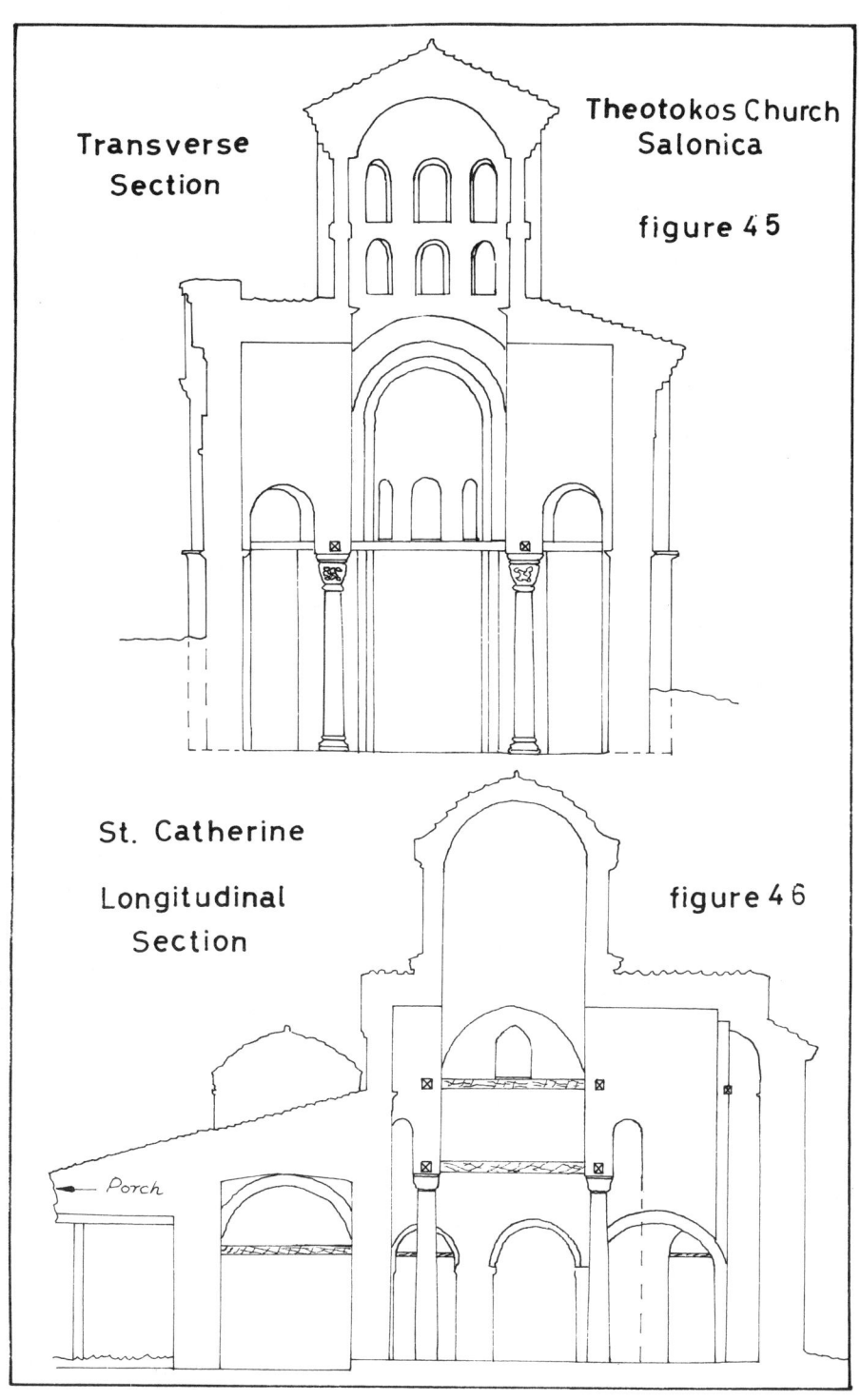

Transverse
Section

Theotokos Church
Salonica

figure 45

St. Catherine

Longitudinal
Section

figure 46

Porch

used but are not always fhreaded through the masonry. Instead, the beams are set a little way into the masonry or wedged against it (fig. 32). Presumably, this makes easier the replacement of the beams if it becomes necessary, but it does make them more obtrusive. The drawing shows the arrangement under the dome.

Further west, the building material was stone instead of brick. The types of churches were more varied and older plans, or modifications of them, continued in use. An example of this is H. Sophia at Ohrid in Yugoslavia, a hall church with three equal aisles. Apart from the intra-mural timbers, already mentioned, tie-beams were used in the north aisle arcade and across the nave and the diaconicon. They can also be seen in both stories of the west front, the exonarthex. This frontage was copied almost exactly in the Fondaco dei Turchi in Venice, built in 1317 with tie-beams which were removed in the 1880 restoration (de Beylie). The portions of H. Sophia with tie-beams were built by Czar Boris between 852 and 889 (Forlati). Hosios Lukas, at Phokis, is a monastery that contains two churches, both of which are of the cross-in-octagon plan, an enlargement of the domed octagon, and both contain tie-beams. The larger church, the Katholikon (*c.*1020), contains timber reinforcement in both the ground floor and the gallery, but the smaller church of the Theotokos (*c.*1040) has reinforcement only in the ground floor. Another cross-in-octagon church is the Pargoritissa at Arta. This building is literally tied together with timber reinforcement below the dome which is considerably higher than most Byzantine domes. The church dates from the end of the thirteenth century (Ebersolt). At Mistra, amongst the upstanding remains of the medieval town on the western side of the River Eurotas opposite Sparta, there are the ruins of several churches, mainly of the cross-in-square type. Five of these contained tie-beams and the earliest of these is the Brontochion built *c.*1300 (Krautheimer). The church resembles H. Irene in Constantinople as it was when it was remodelled in the eighth century so the Greek church is in some degree a renaissance of earlier Byzantine architecture. It will be remembered that the remodelling of H. Irene at that time was mainly the rebuilding of the main vaults without tie-reinforcement, although tie-beams remained in other parts of the church. The architect of the Brontochion copied the tie-beams but used them also in the upper and lower arcades of the nave. The holes into which this reinforcement was fixed still remain. In the atrium of the Metropole (1310), tie-beams were originally used but in some cases were later replaced by tie-rods. The Peribleptos church, built during the first half of the fourteenth century, contained tie-beams under the dome. Very similar evidence to the beam-holes that still exist in this church is found in the springing of the arches beneath the dome of the church of H. Sophia (*c.*1365) in the same town. The columns supporting these arches are tall and slender in both churches. The church of the Pantanassa, built about 1428 (Millet), was constructed on a cross-in-square plan. It contained tie-beams in the aisle arcades, the narthex and in the northern porch. At the church of Asteriou in Attica, beam-holes can be found in the nave arcades. On Mount Athos a group of monasteries was built at dates ranging from the tenth to the nineteenth centuries. Several of these churches contain tie-beams. The Chilandari church (thirteenth century) has arcade reinforcement, the Iveron monastery (1453–60) contains tie-beams in some of the monastic buildings and the church in the Esphigmenou monastery (beginning of the nineteenth century) has wooden reinforcement.

In the Balkans, examples of timber reinforcement can be found at Bačkovo,

Bulgaria, across the entrance to the church built after 1083 (Ebersolt). Tie-beams span the apse in the church of Cisnadioara in Romania. They connect the springers of transverse arches that support a groined vault. The church itself is dated between 1200 and 1225 (Oprescu), and an unusual feature of the tie-beams is their polygonal section. In Bulgaria again, Boyana church has tie-beams across the nave which date from 1259 (Stoikov). Another church at Staro Nagoricane, dating from 1312 or 1313 (Dereko) has wooden tie-beams in the aisles below the dome and in the narthex.

An unusual group of churches is to be found in the Persian town of New Julfa. It was to this place that Shah Abbas the Great transferred a large number of Armenians during the seventeenth century. The churches date from this period and all are of Byzantine or Armenian plan and all contain tie-beams either under the domes or across the naves. About ten good examples are illustrated in Carswell. Late examples of reinforcement are to be seen in the nineteenth century churches of New Trebizond in Turkey (Ballance). Most of them are tied together with iron rods.

Apart from the spread of reinforcement throughout the Christian countries of the Eastern Roman Empire, it was also used in buildings constructed by (or for) the Moslem conquerors of parts of the Middle East. The outstanding building of this sort is the Dome of the Rock (Kubbet es-Sakrah) at Jerusalem (Creswell) probably inspired by the rotunda of the Anastasis at the Church of the Holy Sepulchre. It was built between 685 and 691 and included tie-beams covered with sheets of either copper or bronze worked en repoussé fixed to the beams by large nails. The beams, 75 by 30 cm. (30 by 12 in.), were dovetailed together and laid in pairs above the dosseret of each capital. Above the double beams were laid the springer blocks supporting the arches of the lower arcade. The building is one of the greatest monuments of Moslem architecture, built as it is over the rock where Mohammed is said to have ascended into heaven. Mosques used both tie-beams and tie-rods and one example will have to do duty for hundreds. This is the mosque at Fustat, Egypt, finished in 827 (Creswell), which used tie-beams in the arcades.

Before moving further west and looking at churches which, although Byzantine in plan and construction, share some Western influence, it might be as well to point out what has become increasingly clear in the past pages of this account: that the tie-beam was a very common companion to the cross-in-square church and its variants. Another point that should be emphasized was the enormous popularity of the cross-in-square; so much so that, during the Middle Ages, it must have seemed to Westerners the archetypal Byzantine church. Probably, because of this, it had not much influence on Western design, and one would hesitate to suggest that one of its constructional features, the tie-beam or tie-rod, should have been taken out of its context and used in Western churches by itself. It seems more probable that the tie-beam and tie-rod, previously introduced into the West, as we shall see later, were used in situations suggested by their location in Byzantine churches of the time, amplifying the uses already current in the West. In other words, the example of tie-beam use in cross-in-square churches simply provided further exemplars to Western architects already familiar with the technique.

The following examples of Byzantine churches were not very accessible to western architects and probably had little influence on their building practices in medieval times. The Byzantine churches that had such influence were probably ones in the Venetian area and in southern Italy, not because they were always important instances of fine construction or were famous buildings, but because they were close at hand. No doubt some of the Western architects had the opportunity of seeing great churches of the Eastern Roman Empire, perhaps during the Crusading period or during the Latin occupation, but it can only have been the fortunate few who were able to get so far. To most of them, the mention of H. Sophia and other great churches would conjure up pictures of enormous splendour in descriptions brought back in pilgrims' tales, and not constructional details. However, even these probably had more influence than the churches in Yugoslavia whose description now follows.

In the region now known as Yugoslavia both Western and Eastern influences were important from the twelfth century onwards. The Western influences came from northern Italy, and the Eastern from Byzantium. Such influences could have little real architectural effect on a disjointed tribal situation, even though the tribes were Christian, because the economic resources were lacking. But the foundation of the Serbian state provided the necessary financial stability. The earliest churches are found in the area around the River Ibar, to the north of Macedonia, in the heart of the mountains. In such a position the new state was able to enjoy a good deal of independence and we find that the churches are not simply straightforward copies of buildings elsewhere. It is natural that they are based upon Byzantine plans, but there are many quite un-Eastern details about them.

The earliest churches are usually classed as Raškan after the area where they are found, and most are associated with monasteries. One such monastery is that of Studenica and the church is a cross-in-square design with a dome reinforced by tie-beams set at right-angles below it (fig. 47). It dates from between 1183 and 1191 (Millet). Western influences include decoration of the western doorway and the Romanesque pilasters. Evidence of tie-beams can be seen in the church of Morača built in 1252 (Dereko) where the holes in which they were set remain in the pillars supporting the dome. The dome of the narthex of Sopočani is reinforced with tie-beams and dates from between 1255 and 1276 (Millet). A very similar church is the cathedral at Arilje built between 1290 and 1307 (Millet). Here wooden tie-beams reinforce the dome. Dečani Monastery contains many Italian features including a western doorway and a carved tympanum, but it is basically a Byzantine church. This is confirmed by the presence of the dome and the use of timber reinforcement set at right-angles below it (fig. 48). The church dates from the years between 1327 and 1335.

Other buildings in Serbia, apart from the Raškan group, used timber reinforcement. Gračanica (1321), a cross-in-square church, has a very tall dome resting on strikingly slender pillars which are tied together by two tiers of beams. The building at Parija, near Priboj, built in 1329 (Dereko), contains tie-beams in the western porch. Lešnica church (1341) contains tie-beams (Millet). A few years later, in 1368 (Dereko), Sv. Spas at Prizren was built and the evidence for the use of tie-beams can be seen in the surviving beam-holes. The church at Piača (1358) contains the same kind of evidence (Dereko). An example of the domed square plan can be seen in the church of the monastery of Matka on the River Treska, constructed some time during the fourteenth century (Dereko). This church has tie-beams reinforcing the dome. The dome of the church of the Markov Monastery was treated in the same way and was probably built a few years later in 1371

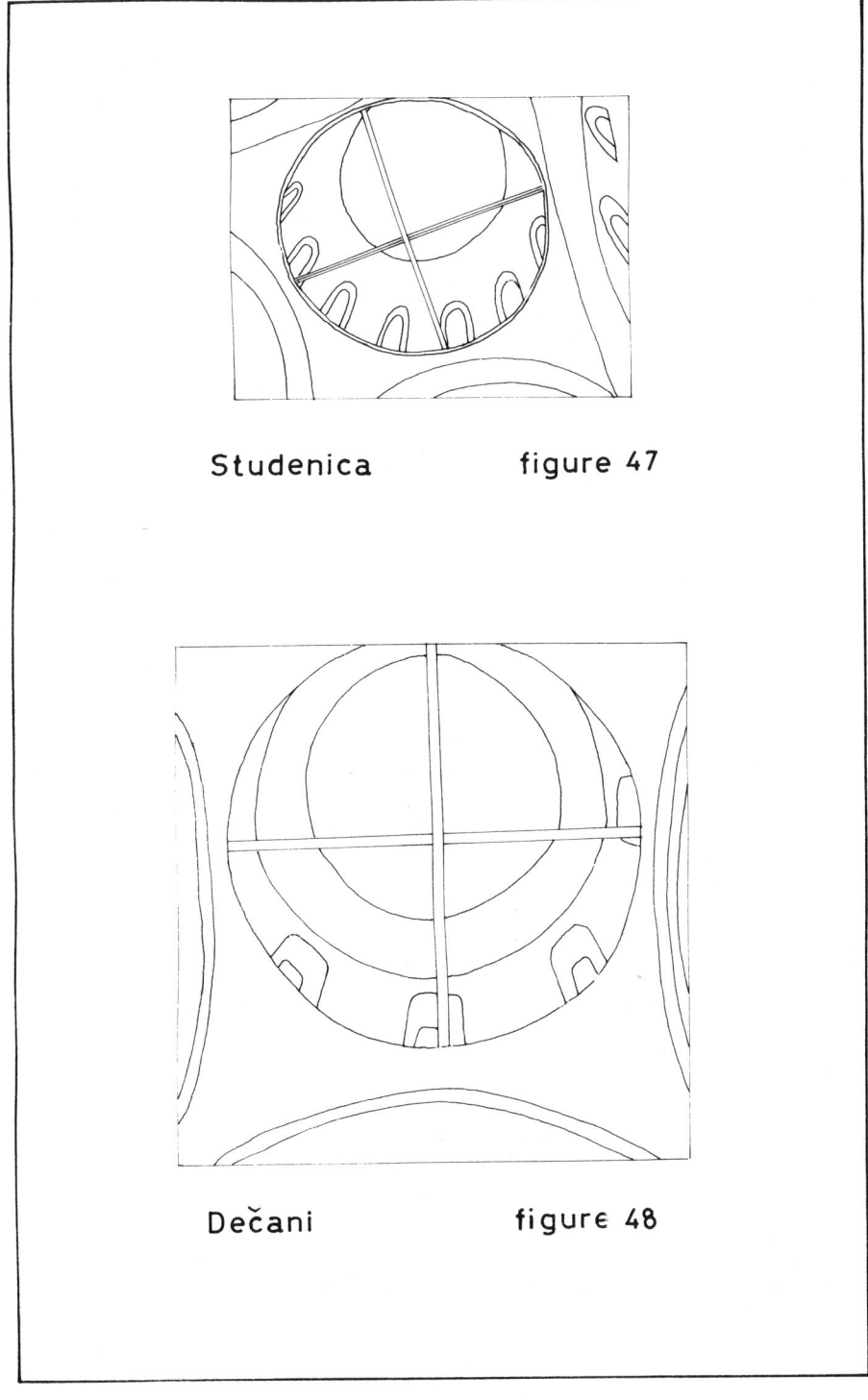

Studenica figure 47

Dečani figure 48

on the cross-in-square plan (Dereko). Another plan that became increasingly popular during the fourteenth century was the trefoiled cross-in-square. Ravanica is an early example (1357–77) in Serbia (Dereko). The pillars beneath the dome are here reinforced with tie-beams. In a church with the same plan, Ljubostina, the same technique was used a few years later in 1385 (Dereko). Later cross-in-square plans can be found at Nova Pavlica (1392), with tie-beams in the usual position between the piers supporting the dome, and at the church of Sv. Arhandjeli at Štip built during the fourteenth century (Dereko). The last few examples were all built on the trefoiled cross-in-square plan. The church of the Mother of God at Kočevište dates from the late fourteenth century (Millet) and has tie-beams in the dome. Another example is the church of Manasija (1418), a church with a lofty dome like Gračanica raised on slender pillars reinforced with tie-beams. Smederevo church is strengthened in the same way and dates from between 1430 and 1456 (Dereko).

An interesting distinction is apparent in Yugoslavia during the medieval period between the Byzantine-plan churches described above and churches of Western inspiration which are tie-rodded. Trogir (1206–55) (fig. 49), Korčula (thirteenth-century) and Sibenik (1431–56) all contain tie-rods and are all Western-style churches. The only exception to this is Sv. Chrysagone at Zadar which is a timber-roofed basilica in the style of similar churches around the head of the Adriatic Sea, and like them it contains wooden reinforcement.

Before we leave the Byzantine-style church, there is one area still left to consider, and this is the area of the south of Italy. Southern Italy was Byzantine before its conquest by the Normans but little remains of churches built at that time. One example which still contains wooden reinforcement is the Baptistery at S. Severina, Ionio, in Calabria. Built of rose-coloured granite, it has columns tied into the apse with large tie-planks. The plan of the building bears some resemblance to that of S. Vitale, Ravenna. The baptistery dates to around the turn of the eighth century. Later churches in the area and in Sicily are very often partly Byzantine in plan and style, but were built under Norman rule and will be dealt with in a later section. Also discussed later are the Byzantine-style churches in Venice and its area.

There still survive in Western Europe several churches built before A.D. 1000 which contain or contained reinforcement. The earliest of these are the Constantinian basilicas of S. Sabina, Rome (421–32) and Pula Cathedral (probably fifth-century). To these can be added Grado built between 571 and 580 (Krautheimer), consisting of a three-aisled basilica with a single apse and a wooden roof. Both arcades are reinforced with tie-beams (pl.8). Another in the same area is the cathedral at Caorle, basically of the fifth to sixth centuries, but restored in 1038 (*L'Architettura Cronache e Storia*). It is of much the same plan as Grado with a similar wooden roof, but columns alternate with piers in the arcades. These arcades are reinforced with timber beams in the same way as those at Grado. A later church in the same style is S. Maria in Sylvis at Sesto al Reghena, Pordenone. It is a timber-roofed basilica built with tie-beams across the arches under the crossing and dates from 762 (*L'Architettura Cronache e Storia*). These churches are in a region at the head of the Adriatic where Byzantine influence was strong and are undoubtedly provincial buildings; the manner of their construction is not inspired, nor were all the materials being used for the first time. The basilica in

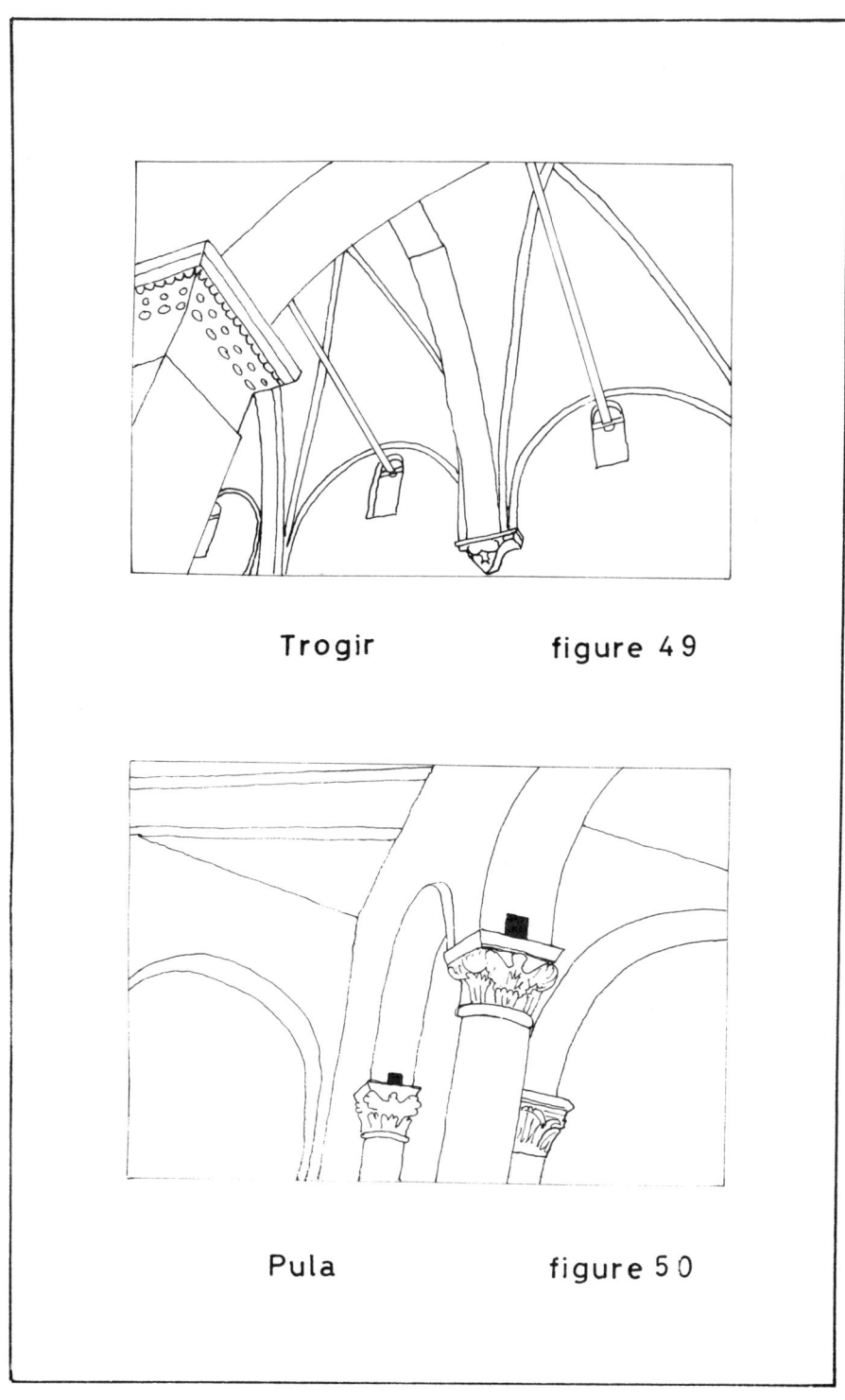

Trogir   figure 49

Pula   figure 50

Rome is of a much higher standard of construction. Its murals and mosaics proclaim that it was built as a grand church and the use of tie-rods instead of tie-beams is further evidence that no expense was spared to use the best, or rather, the most fashionable materials. Two other churches in Italy demonstrate this distinction. The baptistery at S. Severina uses wooden reinforcement and is a provincial building built of local materials on a Byzantine-style plan. S. Satiro, Milan, described by Conant as a perfect four-column church, also built on a Byzantine-style plan, dates from 876 when Milan was more than just a provincial centre. The church was constructed on a splendid scale, fully vaulted with barrel-vaulted nave, barrel vaults north and south of the dome and groin vaults elsewhere. Tie-rods are found throughout.

It is certain that iron and wooden reinforcement were established in Western Europe both in the earlier, basilican-type church and in later Byzantine-plan buildings by the latter part of the ninth century. It is almost certain that these usages were due to Byzantine influence in view of the fact that reinforcement was common in all kinds of churches in the Eastern Mediterranean and does not appear in Western Europe before the year A.D.1000 except in buildings or areas where Eastern influences are suspected. It is assumed that the reason for the use of reinforcement in Italy was the same as that in the Eastern Mediterranean. In parts of Italy earthquakes are reasonably common and were so during the medieval period. Similar precautions as those used in the East could be taken in Italy. Earlier buildings in the Italian peninsula (almost all Roman) did not employ either timber or iron reinforcement. The Roman buildings were constructed as monoliths with roofs as homogeneous concrete units and so were fairly earthquake-proof. It seems reasonable, therefore, to suppose that the introduction of reinforcement and the early use of it was done as a conscious precaution by builders who had experience of Eastern Mediterranean constructional techniques.

The period around A.D.1000 was a strange one in Europe; Byzantine influence was very strong. The Ottonian Holy Roman Emperors were reviving the imperial ideas of Charlenagne: Otto III, the son of the Empress Theophano of Constantinople and a descendant of the House of Saxony, took up residence in Rome and there struck seals announcing his intention: 'Renovatio Imperii Romani'. It was certainly overdue, for not a single building had been erected in Rome for over a hundred years. The architectural awakening with which reinforcement was associated began in Italy and in France simultaneously and in Germany a little later. It has already been pointed out that Byzantine influence was now strong in Italy; the use of reinforcement in churches of the first part of the eleventh century points towards this conclusion. In France, Burgundy saw the beginning of an architectural revival centred on Cluny, and the use of reinforcement in Burgundy must indicate some Byzantine constructional influence (although probably indirect) as it does in Italy at this time. In Burgundy and other areas the function of reinforcement as a protection against earthquake is no longer necessary and the technique is used as a safeguard against structural deformation due to the unequal settlement of the building during the first few years of its life. In many cases the reinforcement was retained indefinitely and remains in place to this day. This change of function must have taken place at this time and in areas outside the earthquake regions of Italy, but the idea of reinforcement was still in essence an Eastern Mediterranean one.

Carolingian architecture does not use reinforcement. In this way it is a true descendant of Ravenna where reinforcement is not found either. Little remains of Ottonian architecture and no churches survive containing reinforcement apart from a late example: the church of St. Maria in Capitol, Cologne, which dates from between 1049 and 1069. This church, before the damage inflicted during the Second World War, contained tie-beams across the arch between the apse and the crossing below the dome and spanning the arch between the nave and the crossing. The system recalls the habitual Byzantine scheme of reinforcing the arches beneath the central dome. The church has no analogies closer to the Rhineland than Italy where its supposed model is the church of S. Lorenzo in Milan, but it was undoubtedly influenced by Byzantine ideas stimulated by Imperial enthusiasm.

In Italy, the period after A.D. 1000 is marked by a great increase in building activity in both the Byzantine-inspired styles and in the native style which is variously described as Romanesque or Lombardic. The Byzantine influence is strongest in southern Italy and Sicily and in the area of northern Italy around Venice. The churches of this time containing timber reinforcement are listed below:

Date	Church	Influence
1007–19	S. Liberatore, Serramonacesca	Romanesque
1008	S. Fosca, Torcello	Byzantine
1008	S. Maria Assunta, Torcello	Byzantine
1021–31	Aquileia Cathedral	some Byzantine
1050–75	S. Benedetto, Valperlano, Lenno	Romanesque
1063–95	S. Abondio, Como	Romanesque
end C11	SS. Pietro e Paolo, Agro	some Byzantine
C11	Rossano Cathedral	Byzantine
1094	S. Nicola Pellegrino, Trani	Romanesque
1089–1132	S. Nicola, Bari	Romanesque
1110	S. Lorenzo, Verona	Romanesque
1117	S. Pietrodi, Legnano	Romanesque
1117	Nonantola Abbey	Romanesque
1118	S. Antimo at Montalcino, Siena	Romanesque

As can be seen, the earlier tie-beamed churches include several which were Byzantine-influenced, but it is clear that timber reinforcement was freely used in the Romanesque churches after 1100. The church in the Benedictine monastery of S. Liberatore was a three-aisled basilica, now in ruins (Gavani). There was a wooden roof and the arches leading into the three apses were strengthened with tie-beams. S. Abondio, Como, a five-aisled basilica, roofed mainly with wood, with five apses, contains a sort of interior narthex (Conant) and this was reinforced with tie-beams, the holes for which still remain. This church embodies Cluniac architectural ideas as a result of its position on one of the routes leading northwards into Burgundy. Another church on the same route is the less important church of S. Benedetto at Valperlano, Lenno, on the western shore of Lake Como. It has a tie-beam supporting the arch which leads into the central apse. It is interesting to note that two out of the three churches of Romanesque type that use tie-beams at this period are on the route to Burgundy and lie very close together. The two churches at Torcello in the Venetian lagoon were built at a time when Torcello was still a rival to the growing city of Venice. The cathedral of S. Maria Assunta is a three-aisled basilica with three apses and is reinforced with tie-beams along the nave arcades and with prop-beams across the aisles from alternate pillars. From the same pillars, tie-beams are placed transversely across the nave (fig. 52). The church was originally built in 641, but was rebuilt twice, in 864 and in 1008. By the side of the cathedral stands the little church of S. Fosca, a Greek-cross octagon with exterior galleries (Franklin), reinforced both inside and out with tie-beams. The interior system of reinforcement is now incomplete but, originally, double tie-beams spanned all the arches which were intended to support a dome that was never completed (pl.30). Other tie-beams joined the columns to the exterior walls (pl.8). In the same area of Italy is Aquileia Cathedral, rebuilt between 1017 and 1031 after the earlier church had been damaged by earthquake. The new church was protected against further similar damage by prop-beams spanning the nave between the walls above the nave arcades, with their ends supported on corbels set into the walls (pl.10).

In the south of Italy, the monastery of Patire, Rossano, was constructed during the eleventh century when the area was under Norman sovereignty (Venditti). The church was built in Byzantine style with tie-beams reinforcing the arches that support the central dome. Another church built under Norman rule during the eleventh century is the church of SS. Pietro e Paolo at Agro, near Messina in Sicily. The church is again of Byzantine plan, cupolaed, with tie-rods and tie-beams between the pillars supporting the main dome. The church is a curious blend of Byzantine, Norman and Sicilian elements. Two churches also in the south, but owing more to Romanesque influences than anything else, are the churches at Bari and Trani in Apulia. The church at Bari was to be built on a grand scale but various difficulties prevented the builders' ideas from being completely realized. The western porch or narthex was never completed, but the springers of the arches intended for it can be seen on the facade together with the holes for timber reinforcement. Beam holes in the arcades of the nave (fig. 61) show that timber reinforcement was thought necessary in some locations, but lack of them transversely across the nave brought about disorders which resulted in the building of diaphragm arches in 1451 (Ricci). Another church built by the new Norman rulers of Apulia is the cathedral of S. Nicola Pellegrino at Trani begun in 1094 (Willemson and Odenthal). It is a building of unusual height, built above the remains of an earlier church, and the nave arcades contain beam-holes for timber reinforcement. The vestibule capitals show the same evidence.

Verona was a thriving city in Lombardy at this time. Its economic prosperity was reflected in its buildings and Verona became an important architectural centre with a style of its own. Porter (1966) has it that the churches there in the twelfth century perpetuate the type of basilica with transverse arches that had developed at Milan in the tenth century. The style was characterized by transverse arches, alternating supports and wooden roofs, a list to which can be added reinforcement. All these features are present in S. Lorenzo, a three-aisled basilica with three apses (White). Double tie-beam holes appear in the nave arcades and beams were set across the nave at the height of the springing of the barrel vault. The transverse tie-beams have been replaced in recent years by tie-rods. Just outside Verona is

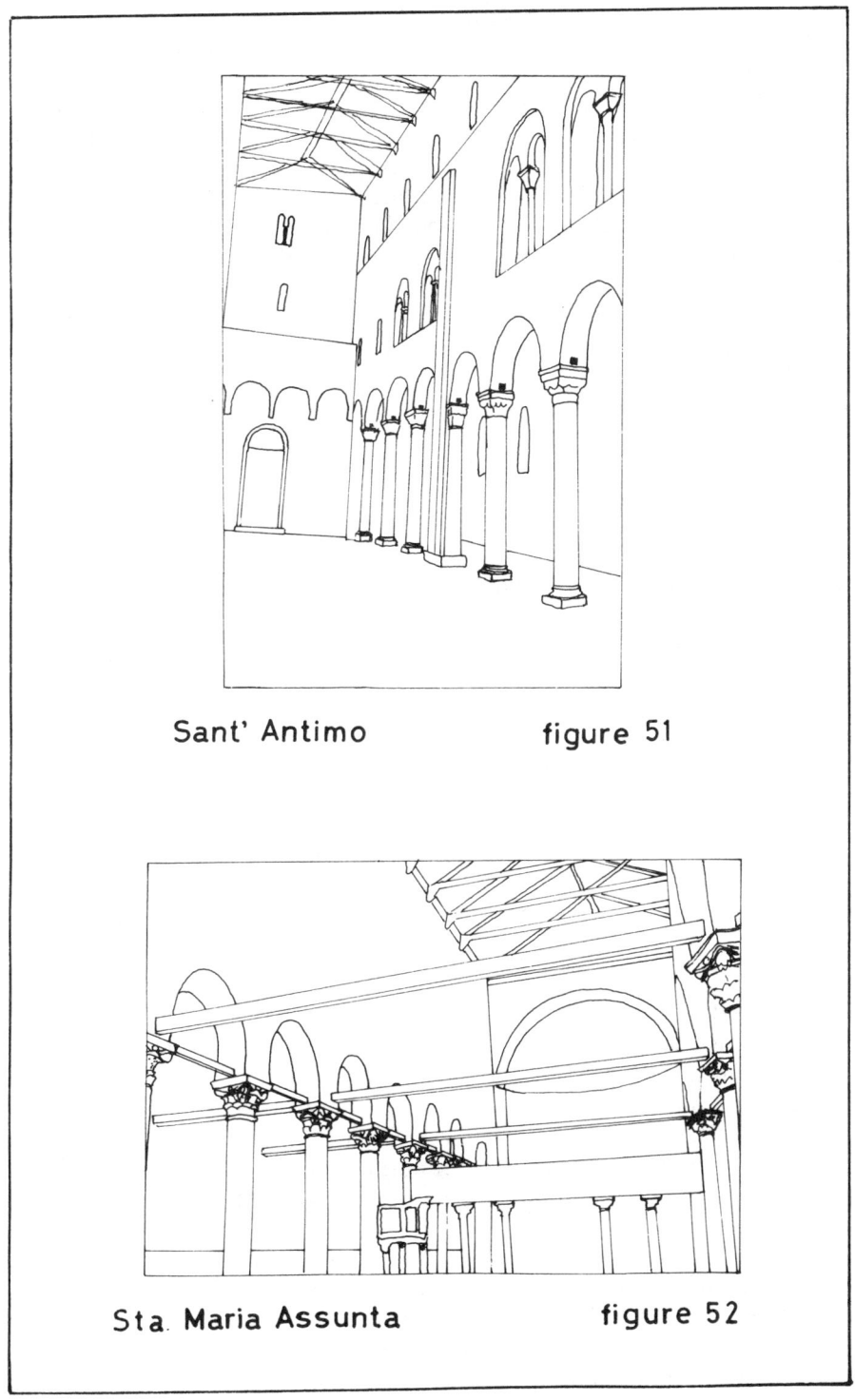

Sant' Antimo   figure 51

Sta. Maria Assunta   figure 52

the small town of S. Pietro di Legnano containing the church of S. Salvaro with double tie-beam holes in each arcade. The church dates from 1117 (Arslan).

Near Modena is Nonantola Abbey, built by 1117 (Bianchi). It is a brick-built three-aisled basilica with a wooden roof. Presumably, the intention was to vault throughout, but this was not carried through, for the nave lobes of the four-lobed columns in the arcades are not carried up above the capitals on which the bases of the transverse arches would have rested. Tie-beam holes occur in both arcades of the nave. Outside the small town of Montalcino in the province of Siena is the church of S. Antimo. The church was much altered in the eighteenth century when the triforium chambers were turned into a number of apartments complete with fireplaces. The wooden-roofed nave, however, remains unchanged. Tie-beams survive in the apse and beam-holes in the nave arcades (fig. 51). The church has an un-Italianate air about it and it would seem possible that there was some influence from north of the Alps.

The list of early tie-rodded churches in Italy is as follows:

Date	Church	Influence
*c.*1000	Trieste Cathedral	Romanesque
1025	S. Maria Maggiore, Lomella	Romanesque
1028–40	S. Carpofore, Como	Romanesque
post-1040	S. Maria, Calvenzano di Vizzolo Predabissi (Milano)	Romanesque
1069	Parish church at Trebbio (Modena)	Romanesque
1080–1100	S. Vitale in S. Stefano, Bologna	Romanesque
after 1060	S. Pier Somaldi, Lucca	Romanesque
*c.*1097	S. Giacomo di Rialto, Venice	Byzantine
1097–1116	S. Fruttuoso, Genoa	Byzantine
C11	S. Maurizio, Canavese	Romanesque
C11	S. Michele degli Scalzi, Pisa	Romanesque
1100–25	S. Martino, Nereto (Abruzzi)	Romanesque
1117	SS. Trinita, Verona	Romanesque
1117	Parma Cathedral	Italian

Trieste Cathedral is a double church, with the church of the Assumption and the church of S. Giusto built side by side. Both contain iron reinforcement and both were built about 1000. It is difficult to say whence the influence that recommended tie-rods came. Certainly not from the area of Yugoslavia to the east. If from the west, the only surviving church from an earlier date (apart from S. Sabina, Rome, which is too early) containing iron reinforcement, S. Satiro in Milan, could be the source, but it would have been indirect, for S. Satiro is built on a Byzantine plan. It might be that the idea for the use of iron came from there, while the general knowledge of reinforcement derived from earlier timber examples in the area around the northern shores of the Adriatic Sea.

Several kilometres south of Milan is the town of Lomella. It was one of the residences of the Lombard kings and important before the main centres of building activity shifted to municipal centres like Milan in the twelfth century. S. Maria Maggiore is a flat-roofed basilica with the central nave covered in timber. It is very much a Lombard design and contains tie-rods used to strengthen dia-phragm arches across the nave. The proximity of Milan indicates a possible inspirational source for the tie-rods. The church is dated by almost unanimous opinion to *c.* 1025. S. Carpoforo in Como is a similar church with a wooden roof and diaphragm arches across the nave. These arches and the nave walls are held together by iron tie-rods. It is probable that the church dates from between 1028 and 1040 (Conant). Another church not far from Milan is S. Maria, Calvenzano di Vizzolo Predabassi. This church was erected in 1040 and was originally roofed in timber with tie-rods set transversely across the nave (Porter, 1915–17). The parish church at Trebbio in the province of Modena contained tie-rods in the nave arcades and across the nave and was built in 1069 (Bianchi). One of the churches that make up the complex of S. Stefano in Bologna is SS. Vitale e Agricola. It has a long domed vault above the nave with transverse arches supporting it. The church was probably originally fully reinforced, but today only some of the tie-rods remain across the aisles and in one arcade. In some places the iron hooks that held the rods in position can be seen. The church originally dates from a very early period, but the present building is the result of a reconstruction between 1080 and 1100 (Conant). Tuscany provides an example during the later eleventh century, after 1060 (Salmi). This is S. Pietro Somaldi, Lucca, and has tie-rods across the nave.

The only two Byzantine-plan churches in the list were built in ports which were engaged in trade with the East, namely Venice and Genoa. In the former, the area of the Rialto Market was being developed during the latter part of the eleventh century. Venice was beginning its climb towards prosperity and the first important buildings were rising. One of these was the church of S. Giacomo di Rialto. The plan came from the East—a Byzantine domed cross-in-square with tie-rods throughout the interior. The church dates from *c.*1097 but has been restored several times since (Salvadori). In Genoa, the church of the abbey of S. Fruttuoso was built on a Byzantine plan between 1097 and 1116 (Ceschi). It is a double church and the upper church has double tie-rods below the cupola.

Further north, in the province of Turin, is the church of S. Maurizio at Canavese. It has tie-rods reinforcing the aisles and was built during the eleventh century (Oliveri). S. Michele degli Scalzi in Pisa has tie-rods across the nave and dates from the same century (Salmi). At Nereto in Abruzzi, the church of S. Martino is a three-aisled apsed basilica. Tie-rods reinforce the aisles and the building dates from between 1100 and 1125 (Gavini). SS. Trinita in Verona, built in 1115 (Porter, 1966), contains tie-rods in the interior narthex. It is probable that the rest of the church was similarly strengthened, but it has been much remodelled and heightened at some period in its history. The cathedral at Parma was the first of the Italian churches to have a high vault over the nave. The church was begun in 1117 (Porter, 1915–17) after a disastrous earthquake that affected a large number of cities in northern Italy. The nave is vaulted with quadriparte and transverse ribs as is the aisle and both vaults are sustained by tie-rods.

All the churches described above are in the north of Italy in the ancient area of Lombardy with the exception of one in Tuscany and one in Abruzzi, both fairly late in the series. This demonstrates the point that the use of tie-rods began in the most advanced and prosperous area where the expense of tie-rods and the neces-sary technical skill was most readily available and where, as we have already seen, the example of S. Satiro at Milan was there to provide the inspiration of iron reinforcement. The use of such reinforcement greatly increased after 1117 when a large part of northern Italy was devastated by an earthquake, an event which emphasized the need for reinforcement. It seems that as many churches as could possibly afford metal reinforcement used it in the general rebuilding, while others had to make do with timber reinforcement. This natural catastrophe, therefore, sharpened the distinction between the more prosperous and the poorer communities—between metal and timber reinforcement. It has already been said that metal reinforcement, on the whole, was inferior to timber, but aesthetic considerations dictated the use of the former. This consideration had fewer deleterious consequences in Italy where most churches depended for the effect on horizontality but, elsewhere, and occasionally in Italy, in buildings where height became an important factor, the use of iron reinforcement proved to be an inadequate precaution against disorders in the structure, whether induced by natural causes or by lack of constructional skill. The prosperity of the north, despite the continual wars between the cities and the quarrels between Pope and Emperor, expressed itself in magnificent building. At this time, as in Venice, civic pride and independence were growing and in their outward manifestation the churches played no unimportant part.

These factors, then, account for the common use of metal reinforcement in northern Italy: the civic pride of the communities, the natural wish to use the latest and most fashionable material, the aesthetic superiority of metal over timber reinforcement and the fear of a recurrence of the earthquake of 1117. It seems reasonable to assume also from the chronology that the widespread use of iron reinforcement began in the north.

Reference has been made to two Byzantine-plan churches which used iron reinforcement during the eleventh century in Venice and Genoa. The location of these two churches is a reminder that these cities were the chief ports of northern Italy during the medieval period with regular and frequent contacts with the Eastern Empire. How much this influence was passed on inland to other builders is difficult to determine, but these two places remained as sites of examples of Eastern technique throughout the period. In Venice, the domed church of S. Maria del Carmelo has tie-rods reinforcing the arcades of the nave. This brick-built structure dates originally from 1125 but has been restored several times since, notably in the fourteenth century (Meldahl). Another church within the sphere of Venice is the church of S. Silvestro at Vicenza built about 1135 (Arslan). It has tie-rods across the nave. In the city of Genoa, for long the commercial rival of Venice in their contacts with the East, there were a number of Byzantine-influenced churches erected during the twelfth century. One of these, built during the early part of that century, is the domed church of S. Lorenzo. The nave is roofed with a barrel vault and is reinforced with tie-rods. The use of tie-rods across the nave is common in churches of Genoa at this period. Another example is S. Damiano, another domed church, built in the mid-century (Ceschi), but with the nave roofed with a groin vault above iron reinforcement. S. Giovanni di Pre, finished in 1180 (Ceschi), has a nave with a ribbed vault with transverse tie-rods. Another kind of vault, a flattened barrel vault, was used in the nave of S. Marco and also strengthened with tie-rods and dates from the latter part of the twelfth century (Ceschi). Despite the variety of roofs, the reinforcement remains standard; it has become part of the building tradition of the area and so remains unchanged by differences of style. This is true of all Italy, but it is interesting to see it demonstrated in microcosm in one place.

A church which might owe something to an eastern influence emanating from Genoa is S. Maria Canale at Tortona, some 70 or 80 kilometres to the north, which dates to *c.*1165 (Porter, 1915–17). The vaulting of the nave consists of a number of domed bays strengthened with tie-rods between transverse arches. A church not far off from Tortona is S. Secondo, Cortazzone d'Asti (Alessandria), dating from *c.*1150 (Porter, 1915–17), which has domical vaults allied to tie-rods. This may also represent influence from Genoa. The same influence can perhaps be detected in S. Bernardo at Vercelli where the nave and aisles have domical vaulting. Both arcades of the nave are reinforced with tie-rods and were built in 1164 (Porter, 1915–17).

Further north still we come into an area dominated by the influence of Milan, but a somewhat independent architectural centre was Pavia. It suffered during the earthquake of 1117 and several churches were built in the succeeding years. Two of these used tie-rods. The earlier is S. Lanfranco with a groin-vaulted nave reinforced with iron reinforcement. It was built *c.*1130 (Porter, 1915–17). The other church is S. Pietro in Ciel d'Oro. It is an interesting building, not least because of its analogies with the Burgundian church at Vézelay (1120–1132). S. Pietro was under construction in 1132 but was not completed until 1182 (Porter, 1915–17). The features it shares with the French church are the groined vault in the nave, a 'narthex' bay at the west end, polychrome masonry in the transverse ribs and tie-rods across the nave. Another church further to the west which was influenced by French ideas is the abbey of Vezzolano at Albugnano which has ribbed vaulting in the nave. It is a Benedictine foundation and was completed in 1189 (Porter, 1915–17). Tie-rods cross the nave. It is possible to see on the exterior how the tie-rods were fixed in position underneath the pilasters before the aisle roofs were built (fig. 53 and pl.29).

In the architectural sphere of Milan is S. Bassanio, Lodi Vecchio, with tie-rods across the nave that may date from 1149, but there is some controversy about the chronology. The abbey at Chiaravallo Milanese is a Cistercian foundation and was apparently established by St Bernard in 1135. The vaults could be as early as 1160, but the church was not dedicated until 1196. Tie-rods span the nave and are fixed across the squinches of the central dome. This is a building which shows dual influence; the French is evident in the general plan, but the dome over the crossing is more evocative of Genoa or S. Satiro in Milan.

Verona has already been mentioned as a centre of architectural influence. In the middle of the twelfth century two churches were built, both using iron reinforcement. One is S. Maria Antica and, apart from its vaulting, is a typical Veronese basilica with three aisles and three apses. Tie-rods span the aisles. The other church is S. Giovanni in Valle, again an aisled basilica in the same simple design as other churches in Verona. The arcades of the naves are strengthened with tie-rods at the western ends. The church is wooden-roofed and dates from 1164 (Arslan).

Outside the spheres of influence of the centres already mentioned are three other churches in the north of Italy. Cremona Cathedral was damaged in the earthquake of 1117 and was rebuilt from 1129 onwards (Porter, 1915–17). Tie-rods were used in the nave arcades (fig. 57). The church of S. Maria Maggiore

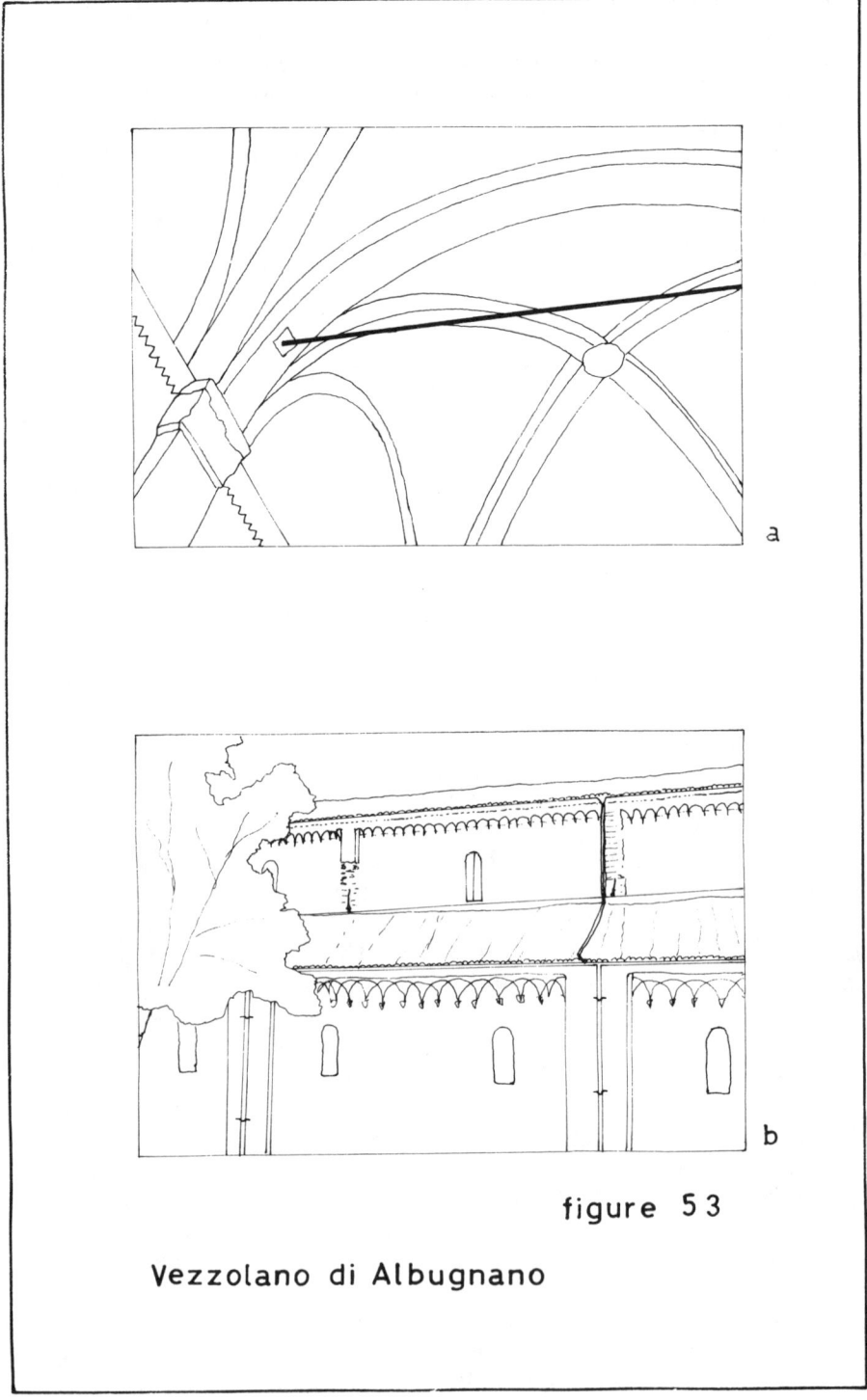

figure 53

Vezzolano di Albugnano

is in the old, walled part of Bergamo and was commenced in 1137 (Porter, 1915–17). The building went on after 1187 and tie-rods were used across the triforium chamber. In Bologna, the previously-mentioned group of churches known as S. Stefano is a red-brick version of the Holy Sepulchre at Jerusalem. Part of it is the Atrio di Pilato dating from 1142 (Conant). The groin-vaulted arcades are supported by transverse arches which are reinforced with metal tie-rods.

The church of S. Pier Somaldi was mentioned earlier as an eleventh century example of tie-rod reinforcement in Tuscany. There are five other examples in the same province during the twelfth century. The church at S. Gimignano, consecrated in 1148 (Salmi), has tie-rods strengthening the nave. At Chiusi, S. Secondiano (formerly S. Mustiola) has tie-rods across the aisles. The church belongs to the twelfth century, but it is impossible to pin down the exact construction date during that century. This is the case also with the church of S. Giovanni in Lucca which also has tie-rods across the aisles and in the arcades of the nave at the eastern end. S. Michele in the same city has tie-rods across the nave and was completed in 1143 (Salmi). The fifth example is the parish church at Arezzo which has tie-rods across the nave at the crossing. The church was destroyed in 1111 and the present building is a reconstruction dating from the latter part of the twelfth century (Salmi).

Further east in Italy, the city of Atri (Abruzzi) contains the Benedictine abbey of S. Maria built, possibly under French inspiration, as a groin-vaulted, aisled basilica (Gavani). The transverse arches are reinforced with tie-rods that span the nave. The church was erected between 1175 and 1200.

Wooden reinforcement after 1117 was rarely favoured in the earthquake-shaken north of Italy. As we have seen, iron was very commonly used and, when it is added that only four examples of wooden reinforcement can be cited in the north outside Verona, whose traditions were peculiar to itself, it can be seen that the phrase 'almost universally used' could be applied to iron reinforcement in Lombardy after the great earthquake. Two Eastern-plan churches are both in Veneto. One is on the island of Murano in the Venetian lagoon. It is the church of SS. Maria e Donato, finished in 1140 and restored by a vandal's hand between 1858 and 1873 (Salvadori). It is a centrally-planned church and contains tie-beams both in the nave arcades and beneath the crossing. The small town of Concordia contains a cathedral with a campanile and a baptistery (Schaffen). The baptistery is a small trefoiled-square with tie-beam holes on each side of the arches leading into each lobe. Apart from the beam-holes, the building is identical with two churches across the Adriatic in Yugoslavia. One is Sv. Nicholas near Nin and is now crowned with a small military tower instead of a dome. The other is the tiny, thirty-foot long cathedral of Sv. Croce in the village of Nin itself. They are said by Yugoslav archaeologists to be derived from Croatian wooden architecture, but it seems more reasonable to suppose that Byzantine influence is responsible, for it is unlikely that a Croatian church would have been constructed as a baptistery in Veneto. The baptistery at Concordia is assigned to the twelfth century, but there is really no means of dating it.

Wooden reinforcement in churches of the Veronese tradition is found during the twelfth century after the great earthquake. One of these is S. Stefano, in the city, a church which was reconstructed during the twelfth century (Conant). An examination of the structure leads to the conclusion that the reconstruction mainly involved the raising of the nave walls. This necessitated the use of relieving arches, the top parts of which were left open and the bulk of which were filled with masonry pierced with small, round-headed windows. Through these windows the ends of tie-beams projected on each side of the church where their ends were presumably held together with exterior strapping (pl.11). The same device can be seen at the church of the Pantanassa, Mistra, Greece (early fifteenth-century), at Bussy-le-Grand, Burgundy, probably of the sixteenth century, both timber examples, and a metal example at Trogir Cathedral, Yugoslavia (pl.12) in the first part of the thirteenth century. Later remodelling involved the removal of these tie-beams at S. Stefano, but their ends in some cases are left projecting through the windows. In the body of the church, tie-beams can be found across the aisles. S. Maria in Organo in the same city is a Romanesque church heavily redecorated during the Baroque period (pl.13). In the nave arcades tie-beams still remain, looking rather incongruous in their utilitarian role against the ornate background. In its Romanesque form the church probably dates from 1131. A tiny church which takes the Veronese simplicity to its extreme form is the church of S. Zeno at Castelletto di Brenzone in the province of Verona. It contains a single arcade which divides the church longitudinally into two and supports the ridge of the roof. One plain and two Corinthian-type capitals are incorporated in this arcade and they were connected by a double-tier of small tie-beams. According to Porter (1915–17) it is possible that S. Zeno was built *c.*1135.

Another church in S. Stefano, Bologna, is S. Sepolcro, a rotunda whose pillars are reinforced throughout with tie-beams (pl.14). The church was built *c.* 1150, but has been restored (Conant). This restoration included the replacement of the tie-beams. Today they shine like new wood through a coating of creosote, but they have been in position for almost a hundred years. Looking at them, one can understand why it is that some tie-beams have lasted for fourteen hundred years. In the west and within the orbit of Genoa is Cuneo in whose neighbourhood a domed church was built during the twelfth century (Olivero). Both the cupola and the arcades of the nave are reinforced with tie-beams. This is the church of S. Constanze sul Monte, a few miles north of Cuneo.

The examples of wooden reinforcement south of Lombardy are all in churches which have no obvious Byzantine or Eastern affinities. In Tuscany, the city of Tuscania contains several interesting churches, two of them Romanesque structures originally with tie-beams. The earlier of these is S. Pietro, with evidence of tie-beam reinforcement in the shape of holes in the springers of the northern arcade and across the aisles (fig. 54). There is some controversy about the dating of this monument but, unless one can claim that it belongs to the same period as the basilicas of S. Maria delle Grazie at Grado and Poreč Cathedral (both sixth-century), one must assign it to the twelfth century, at the same time as a similar but less ornate wooden-roofed basilica erected at Zadar in Yugoslavia. This is the church of Sv. Chrysagone built by 1175 (Conant). It is similar in plan and construction to S. Pietro and is fairly obviously derived from Lombardic sources. It has tie-beam holes in the nave arcades (pl.15) and beams spanned the aisles (pl. 16): a similar scheme to that of S. Pietro. One should probably link the Tuscan church with Lombardic influences also, but a good deal later than is usually supposed. The other Romanesque church in Tuscania with timber reinforcement is S. Maria Maggiore, built on the same plan as S. Pietro and also with beam-holes

S. Pietro   figure 54

Tuscania

Sta Maria Maggiore   figure 55

in the nave arcades (fig. 55). It, too, probably owes a good deal to Lombard inspiration and was finished in 1206 (Rivoira).

On the eastern coast of the peninsula, the city of Osimo has a local church which was once the cathedral of S. Leopardo which was rebuilt *c*. 1200. It has a wooden roof supported by very thick piers at the crossing. In these piers are beam-holes used for a system of timber reinforcement consisting of a quadripartite tier of beams transversely across the nave and a single-beam tier longitudinally.

S. Lorenzo, Vicenza, was built in the thirteenth century, probably after 1281, and contains tie-beams across the nave and across the aisles (fig. 58). In Verona, the church of S. Anastasia is a Dominican foundation begun just after the middle of the thirteenth century. It has been elaborately redecorated in the Baroque style. The church has a complete system of tie-beams across the nave, in the nave arcades, across the aisles and reinforcing both the crossing and the apse arches. This system of reinforcement was used in Gothic-style churches of this sort particularly in Venice during the next century, but this church in Verona is the earliest surviving example. We return to Tuscany to the province of Siena for the next church. It is S. Antimo, Castelnuovo dell'Abate. Built in 1239 (Salmi) in the Romanesque style, it has tie-beams in the naves and across the ambulatory.

During the thirteenth century, the proportion of wooden reinforcement to iron sank from the earlier ratio of 1:1 to 1:2. This probably reflects the increasing prosperity of communities in the peninsula. But, in Venice, whose links with the East were increasing, the use of tie-beams continued to be a popular technique. It may be that the architects there realized the superiority of timber over metal. Later examples of timber reinforcement are all in the area of Venetia. At Spilimbergo Cathedral, a three-aisled basilica of simple design, the interior is reinforced with decorated tie-beams throughout the nave arcades, across the nave, across the chancel arch and across the arches of the side apses (fig. 56). The beams are prop-beams, butted up against the piers and held in position by spikes set in the wide abaci. The church was built between 1284 and 1400 (*L'Architettura Cronache e Storia,* 1965/6). In Venice, the church of S. Maria dei Frari was built by the Franciscans. It has tie-beams across the nave, across the aisles and along the nave arcades (fig. 59). It was built in 1340. S. Stefano in the same city has similar tie-beams across the nave, across the aisles along the nave arcades. This church was built by 1374 (Salvadori). Another church with reinforcement is S. Maria dei Miracoli, built between 1481 and 1489 (Meldahl). In this city the use of tie-beams in sophisticated churches such as these persisted as we have seen into the fifteenth century and even when tie-rods were used, in one instance at least they were disguised as tie-beams. This is in the Dominican church of SS. Giovanni e Paolo where decorated boxes were used to encase the tie-rods in the same way as in H. Sophia in Constantinople. SS. Giovanni e Paolo was built in the tradition of churches like S. Lorenzo in Vicenza and S. Anastasia in Verona.

Metal reinforcement during the thirteenth century in the north of Italy is well scattered. To the west, the church of S. Antonio, Ranversa, built at the turn of the twelfth and thirteenth centuries, is Gothic rather than Romanesque in style. Tie-rods reinforce the nave and its ribbed vault. In the north-west is Trento Cathedral with tie-rods in the nave arcades. It is also largely Gothic and took almost a century to build, from 1212 to 1309. S. Pietro, Modena (fig. 60), has tie-rods beneath a ribbed vault and dates from 1206 (Quintavalle). In Venice, the

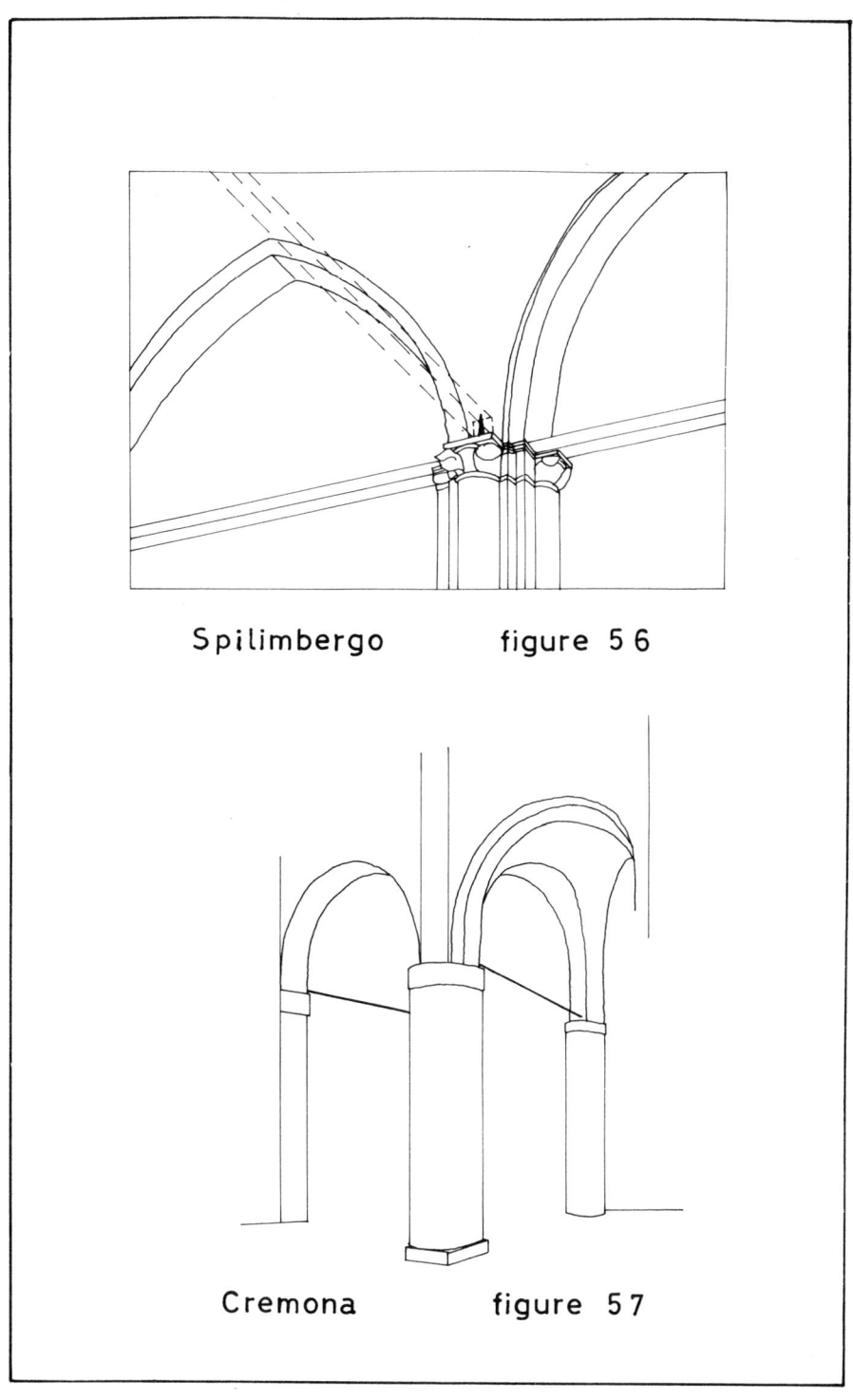

Spilimbergo   figure 56

Cremona   figure 57

S. Lorenzo   figure 58

Sta. Maria dei Frari   figure 59

Modena
figure 60

S. Nicola, Bari   figure 61

church of S. Giacomo dall'Orio contains tie-rods and was built during the thirteenth century from 1225 (Salvadori). Tie-rods across the nave are found in the church of S. Francesco, Bologna, founded in 1236 and dedicated in 1250 (White).

Outside the north of Italy, S. Francesco at Assisi contains tie-rods and dates from about 1228 (Corroyer). Siena Cathedral has some iron reinforcement and was built between 1226 and 1276. Further south, the cathedral at Carrara, erected during the thirteenth century, contains iron reinforcement across the aisles. Lucca has a group of churches built during the same century containing metal reinforcement: S. Cassiano di Conestrone, S. Cristoforo and Villabasilica.

There are many instances of later reinforcement and it will be convenient just to list a selection of them from different parts of Italy:

S. Maria, Propezzano, Abruzzi. Fourteenth century (Gavani).
S. Maria, Ponte Fontecchio, Abruzzi. Fifteenth century (Gavani).
Massa Marittima, Tuscany. Thirteenth/fourteenth centuries (Salmi).
Poggibonsi (Dintorni), Tuscany. Fourteenth century (Salmi).
Florence Cathedral, Tuscany. 1365–1420 (Franklin).
S. Guiseppe, Palermo, Sicily. Seventeenth century.
S. Maria Maggiore, Milan. Fifteenth century (Franklin).
S. Ambrogio, Milan. Reconstructed 1863.
S. Babila, Milan. Modern reinforcement.
Modena Cathedral. Modern reinforcement.
S. Maria in Foro (Chiese dei Servi), Vicenza. 1320 (Arslan).
S. Zaccaria, Venice. 1440–1500 (Meldahl).
S. Maria Formosa, Venice. 1492 (Meldahl).

It was said earlier that the use of reinforcement was known in Italy, Burgundy and Germany soon after the year A.D.1000. It has been suggested too that the knowledge of the technique must have travelled from Italy where the use of it dates back at least to the sixth century. The question of the manner of transmission from Italy northwards is very difficult to answer except to suggest that, in the case of Burgundy, it could have been in a monastic context and, in the case of Germany, as a result of the Ottonian interest in southern building techniques. The evidence for this is purely circumstantial. In Burgundy, two of the earliest churches concerned in this transmission are monastic; of what kind were the buildings at the origin of the lines of transmission in Italy we are not in a position to know. All we can say is that they were probably monastic and were buildings that contained tie-beams. The field is fairly narrow but, without the evidence of buildings no longer in existence, does not provide any firm indication, particularly as the earliest buildings in which the tie-beams may have made their first appearance in Burgundy have almost completely disappeared as e.g. St. Bénigne, Dijon. The earliest surviving reinforced building in Burgundy (Farges) has no obvious parallel in Italy, nor has the major surviving building (St. Philibert, Tournus). All that can be said, in general terms, is that a monastic building like S. Liberatore alla Maietto, Serramonesca, Abruzzi, containing tie-beams originally, which had some influence on Cluny III, might be the sort of building from which the idea of reinforcement came to Burgundy.

Another pointer to the importance of influence from the south is often quoted: William of Volpiano, who was abbot of St. Bénigne at the time when the church was being built, came originally from Italy and might have sent for masons from his homeland when he came to build the abbey church.

These are both rather tenuous pieces of evidence in favour of a southern source for the reinforcement technique but, taken in conjunction with the presence of examples in Italy before A.D.1000 and the complete lack of them elsewhere in Western Europe, it is perhaps enough to be convincing. There is no other reasonable hypothesis.

At this stage it might be useful to sum up briefly what has been brought out earlier in this account. The idea of reinforcement, whether of wood or iron, was not indigenous to Europe. A few examples exist in the period before A.D.1000 between the sixth and ninth centuries, all either in the north or north-eastern areas of Italy—areas open to Byzantine influence intermittently during those years. In the early eleventh century a number of examples are connected with the Romanesque movement and this argues for Byzantine influence on that particular building activity. It is evident during a recrudescence of architecture which also occurred at the same time in the East. But reinforcement is only found in certain buildings, and not in others of the same period and much the same style. The smaller buildings of the First Romanesque did not need reinforcement; it is really only in the larger buildings that it is found during the early years, but not in all of these. We can say then that reinforcement was taken into the structural repertoire of some of the Romanesque/Lombardic masons. In this context it spread across Italy.

One can assume either that this Byzantine influence was re-introduced into Italy from overseas round about the year A.D.1000 or that the technique was generally used in Italy from the sixth century (when it first appears outside Rome) onwards in buildings which have disappeared. The latter hypothesis is possible but not very plausible, for the remaining earliest examples lie in a definite group at the head of the Adriatic where Byzantine influence was strong. Elsewhere, no examples survive, apart from S. Satiro in Milan and the baptistery of S. Severino in the south of Italy. If the technique had been general during these earlier years, one would expect to find a scattering throughout the peninsula. The evidence is on the side of a re-introduction at about A.D.1000 when Byzantine influence strengthened in Italy. At this time we know of a number of reinforced churches in the East from which the idea of reinforcement could have come. The technique was certainly part of the living tradition of Byzantine architecture at that time. The re-introduction probably occurred at more than one place. One such place would be the Venetian area where direct Byzantine influence would be likely. The area around Genoa could be another and there is a definite re-establishment of the technique associated with the churches of Apulia due, perhaps, to the energetic encouragement of Norman patrons.

It seems reasonable to suppose that the principle of earthquake protection (in areas where it was necessary) in buildings directly influenced by Byzantine architecture would have played a part, but that in buildings in areas where earthquake protection was not necessary (where reinforcement was being adapted to other purposes), the Byzantine influence was probably indirect. Thus, one can accept a Byzantine (or Byzantine-trained) hand in some churches of Italy, but not further north in France and Germany. The question of design is important and one must take into account definite Byzantine plans like those of Venice. Others, which have a plan nearer to, or within the sphere of, the Romanesque tradition, must have had a non-Byzantine architect, but one who was acquainted with Byzantine work and who had the genius to adapt one of their important safety precautions to other uses.

Burgundy was a region not much influenced by outside ideas after the early years, until the coming of the northern Gothic, but it was an area that had great influence on other places. The earliest examples of reinforcement in Burgundian churches are listed here. With them are some churches in immediately adjoining regions which, for all practical purposes, were part of the Burgundian architectural scene (fig. 62).

Date	Church	Reinforcement
1001–18	St. Bénigne, Dijon (possible example)	?
1000–25	Farges	Timber
1020–66	St. Philibert, Tournus	Timber
1059–1107	La Charité-sur-Loire	Timber
late-C11	Montiérender Abbey	Timber
1120–32	La Madeleine, Vézelay	Iron
1120–32	Autun Cathedral	Iron
1139–55	Fontenay Abbey	Iron
first half C12	Bussy-le-Grand	Timber
1170–5	Flavigny	Iron

The movement from Italy has been suggested. The other possibility is a direct link with the eastern Mediterranean, but this is difficult to envisage, particularly as the buildings are in a reforming monastic context whose links with the East would hardly have existed at the time and which would certainly have owed nothing in religious inspiration to Byzantium.

St. Bénigne at Dijon (Côte-d'Or) was rebuilt (1001–18) during the abbacy of William of Volpiano who came from northern Italy. One can believe the report that Italian workmen and an Italian-trained architect were employed in the construction, particularly if it was the result of disparate influences described as Roman, Romano-Ravennate and Byzantine-Ravennate by Rivoira. There is no firm evidence that reinforcement existed at St. Bénigne apart from a suggestion in a drawing in Plancher where there are indications above the capitals that might be tie-beam holes (pl.17). However, one must postulate the existence of tie-beams in the area at about that time in a church large and important enough to shed a constructional influence about it. The reason for this is that there is a part of the church of St. Philibert, Tournus, the upper storey of the narthex (1020), and a small church at Farges, built before 1025, which contain tie-beams. A totally new technique would not have been imported into the area simply to be incorporated into a portion of St. Philibert and a village church. The builders of these structures almost certainly got the idea from a famous building not far off, and St. Bénigne would have been an obvious exemplar. Cluny II, another important early Burgundian building, finished before A.D.1000, is too early to have had reinforcement.

The nave of the church of St. Philibert might have stood as an example if it were

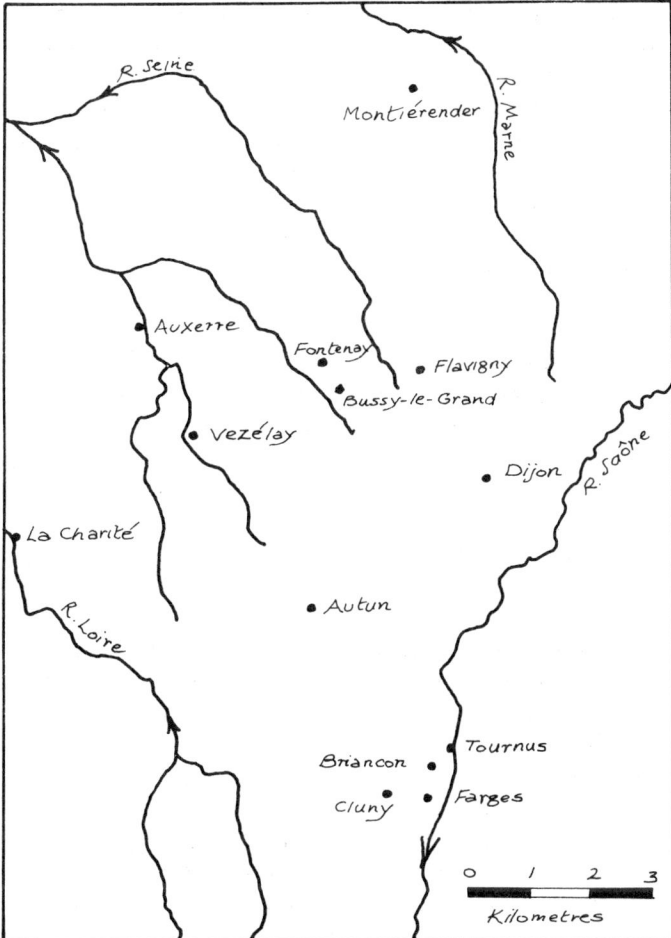

FIG. 62. Map of the Burgundian area

early enough, for it had tie-beams, and is grand enough. The narthex, with tie-beams of the same sort as those at Farges, is not grand enough to have been the source of inspiration in itself, and must have occupied a secondary position relevant to another monument, as Farges does. The building is contemporary with Farges. St. Philibert is thought by various authorities to be the result of influences from different quarters, including Lombardy and the west of France, and it is difficult to draw from these hypotheses any firm general conclusions about the inspirational source of the architectural composition.

It is not assumed that Plancher, in his drawing of St. Bénigne, inserted the indications which might be those of beam-holes, realizing what they were as he drew them, but that he included them incidentally in his section. Presumably, if these beams did exist, they were arranged as they are in the rotunda at S. Sepolcro, in S. Stefano, Bologna. On the analogy of Farges, one would expect tie-beams across the nave as well but, with the destruction of a good deal of the church before Plancher's examination, there is no means of knowing whether this was so or not.

The church of St. Barthélemy at Farges (Saône-et-Loire) was built between

1000 and 1025. It bears the marks of what are usually described as Lombardic traits; it is built of small stones and has Lombardic bands under round arches. Tie-beams across the barrel nave have mostly disappeared, but stubs remain in some of the beam-holes (pl.18). The position of these beams is analogous to those in the narthex of St. Philibert. Other beam-holes exist each side of the chancel arch (pl.19). An interesting feature of the church is the vaulting of the aisles. They are covered with modified transverse barrel vaults which are similar to the vaults above the nave of St. Philibert. St. Philibert, Tournus (Saône-et-Loire) has tie-beams reinforcing the barrel vault of the upper narthex which probably dates from about 1020 (pl.20). A fire destroyed the bulk of the church in 1007–8 leaving the monks to face the prospect of an almost complete reconstruction. Apparently, the ground floor of the narthex still stood, and a gradual rebuilding took place, starting with the upper storey of the narthex and ending at the east in 1120. It seems likely that the style of architecture grew grander as the rebuilding went on. In the nave, tie-beams were used to reinforce the arcades and were set transversely across the aisles. Pl.21 was taken during the last century before these tie-beams were removed, while pl.23 is a recent view of the upper part of the nave piers and demonstrates how almost all evidence of the use of reinforcement can disappear in restoration. The nave is vaulted with transverse barrel vaults and dates from about 1066.

A little way outside Burgundy proper, but still within the zone of its influence is the church of La Charité-sur-Loire (Nièvre). The nave had a barrel vault supported by reinforced transverse arches. The holes for the reinforcement still exist in the remains of the nave (pl.22) and date from between 1059 and 1107. The nave has been partially destroyed and only four bays were rebuilt during the seventeenth century. A church not far away from the Burgundian architectural heartland is the abbey of Montiérender (Haut-Marne). Tie-beams cross the nave at the springing of the high barrel vault. The church dates from the second half of the eleventh century and is an interesting building in many ways. It is the equivalent in France of the great Ottonian churches of Germany and consists of an aisled hall with galleries and clerestories like St.-Remi at Reims. The German churches, however, apart from St. Maria im Kapitol, Cologne, do not contain reinforcement. Another abbey church of much the same period is St. Pierre at Airvault (Deux-Sèvres) which has a nave reinforced with iron and was consecrated in 1100 (Aubert, 1966). What is difficult to understand is the inspiration for the tie-rods at Airvault. They are the earliest in France, but occur in a church that has been restored a good deal since, so that the appearance of the nave belies the date of its original construction. However, there seems to be no reason to doubt the date of the tie-rods. It seems probable that the original inspiration came from Italy.

The builders of La Madeleine at Vézelay (Yonne) probably sought an architectural theme outside the Cluniac Order, and it is likely that this inspiration came from northern Italy and included the notion of tie-rods rather than tie-beams. The tie-rods were used across the nave at the springing of the transverse arches. Two methods of embedding the supporting hooks into the nave walls were used (fig. 4). However, these tie-rods were ineffective and the walls of the nave bulged apart under the weight of the groined vault which was one of the earliest vaults of this size, and the transverse arches were distorted into the shapes they assume today. It became necessary to support the walls with flying buttresses and the tie-rods were dispensed with. The hooks to which the tie-rods were secured still exist in the western part of the nave. Like Anzy-le-Duc (Saône-et-Loire), Vézelay is a fore-runner of the local Burgundian 'half-gothic' style which develops later into the international Gothic style of the Cistercians. The nave at Vézelay was built between 1120 and 1132 (Salet). Another church built between the same dates is the cathedral of St. Lazaire at Autun (Saône-et-Loire) which was also reinforced with tie-rods. The reinforcement was put, not across the nave as at Vézelay, but across the aisles and across the choir. As at Vézelay, the hooks which held the ends of the tie-rods remain.

The use of tie-rods was taken up by builders of other Orders apart from the Cluniacs. At Fontenay (Côte-d'Or), the abbey church is part of the oldest surviving Cistercian ensemble. The pointed barrel vault has arcades reinforced with tie-rods. The hooks still remain. Fontenay is another example of the Burgundian 'half-gothic' and was built between 1139 and 1147.

Bussy-le-Grand (Côte-d'Or) is a village church, originally built with timber reinforcement across the transverse arches supporting a nave which is covered with a pointed barrel vault. In course of time, the beams rotted away or were removed with the result that the weight of the vaulting began to push the walls outwards. To correct this, beams were set across the nave, through the windows and were secured on the outside (pl.24). On the outside of the church, the nave walls were refaced covering both the windows and the ends of the beams. This arrangement is similar to the arrangement at S. Stefano, Verona. The church was built in the first half of the twelfth century (Von Veltheim). Later in the same century, the church of St. Pierre at Flavigny (Côte-d'Or) is another example of the Burgundian 'half-gothic' and has tie-rod reinforcement. The tie-rods strengthen the narthex and date from between 1170 and 1175 (Branner).

The pattern then in Burgundy is of timber reinforcement used initially, giving way in the larger churches to iron, while timber was still being used in the smaller churches. This is the same distinction as was evident in Italy but, as we shall see later, this does not hold good further north in France. It has been said that the use of reinforcement was common in the churches of Cluniac monasteries. This is not entirely true, for only a minority of Cluniac churches use it. What seems to be the case is that the Cluniac Order adopted the technique earlier and more enthusiastically than any other builders.

By the beginning of the twelfth century, timber reinforcement was fairly established in France with examples in Burgundy, in the south (St. Trophime, Arles), in Corrèze (the Cluniac abbey church at Beaulieu-sur-Dordogne (fig. 64)) and in Brittany (Notre Dame at Quimper). The group of domed western French churches did not commonly use tie-reinforcement. One would expect to find the technique used to strengthen the domes as was so commonly done in the eastern Mediterranean but, actually, only two churches have evidence for the use of tie-beams. One is the cathedral of Angoulême (Charente) built between 1100 and 1128 (Aubert, 1966) where beam-holes appear in the lower part of the dome (fig. 63) and the other church is at Solignac (Haut-Vienne) with reinforcement across the arch leading into the northern apse. The church dates from 1143 (Baum).

During the twelfth century, the technique of reinforcement spreads across the country and beyond its borders. The now-destroyed church of St. Etienne, Troyes, was begun in 1157 (Branner, 1960). One of the few remains of this building is a

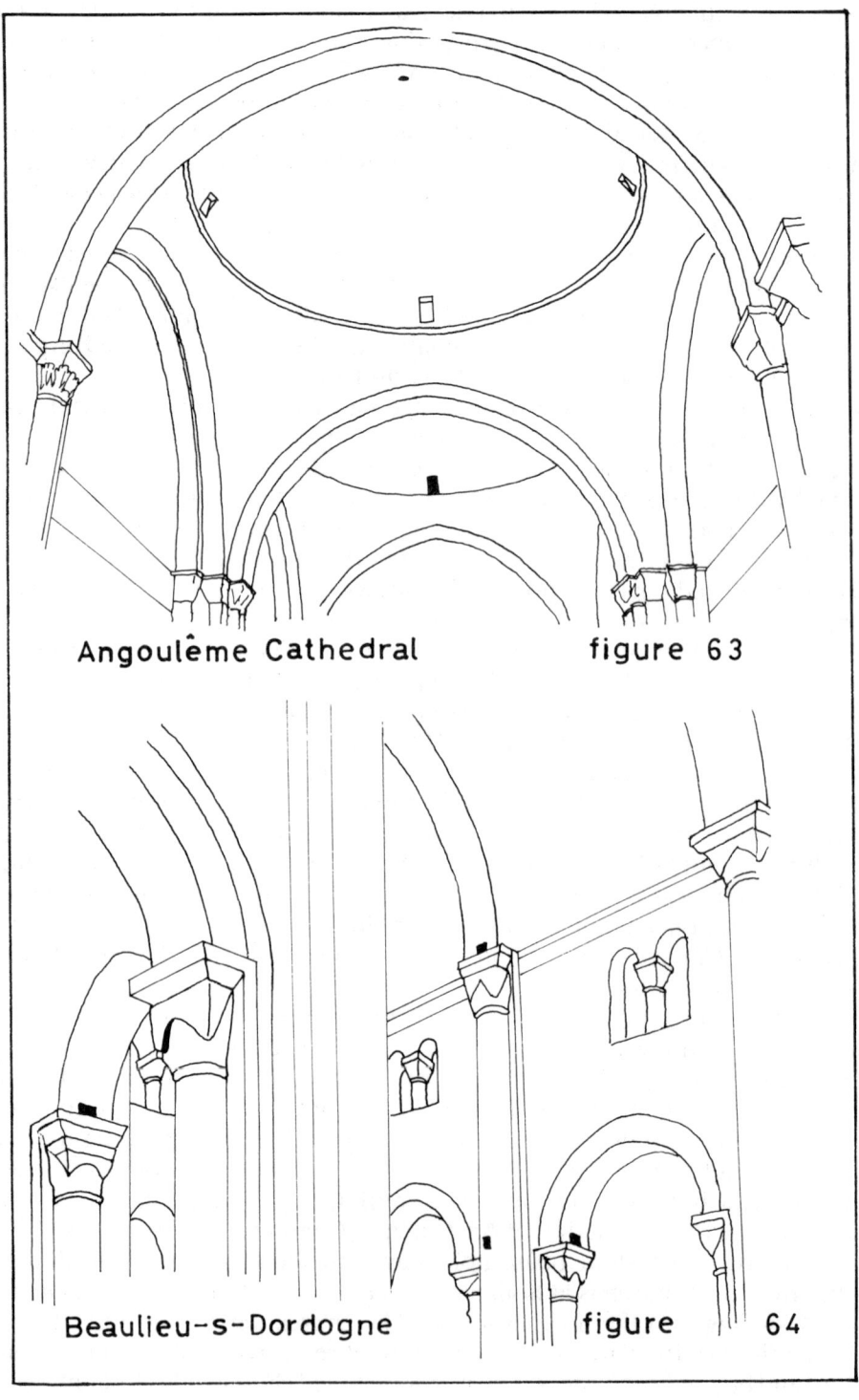

Angoulême Cathedral   figure 63

Beaulieu-s-Dordogne   figure   64

capital which preserves a beam-hole into which a tie-beam was fixed (Musée de la Ville, St. Loup). Just across the Spanish border from Luchon is the small town of Bosost. Its church dates from the mid twelfth century and has holes which contained tie-beams that spanned the nave and the aisles. A quadripartite beam reinforcement existed across the chancel arch. Tie-beam holes and corbels for supporting the ends of prop-beams can be seen at the church of Brancion (Saône-et-Loire), dating from the second half of the twelfth century. The barrel vault is spanned by transverse arches and was reinforced by tie-beams in the three bays nearest the west door (pl.25) and by prop-beams in the eastern two bays. It is curious that the builders thought that the forces created by the vault should be acting in diametrically opposite directions in each half of the nave. The builders' caution can be noted in the size of the windows; tiny, single-splay openings set in thick walls.

In the Vosges, in an area much affected by Burgundian ideas, is the church of Notre Dame at Saint-Dié. It contained tie-beams and dates from the end of the twelfth century (Durand, 1913). This church is one of a group with similar features which include Tournai Cathedral and a number of churches in the Rhineland: Holy Apostles and St. Maria im Kapitol in Cologne, Speyer, Worms, Maria Laach, St. Quirinus at Neuss and Rosheim (Bas-Rhin). All these have reinforcement in one form or another. Also in the east, beyond the borders of present-day France, the church of Notre Dame at Valère, near Sion (Valais) in Switzerland, has tie-beams and tie-rods reinforcing the nave. The tie-rods may be modern. The church dates from c.1200 (Enlart, 1927). A very late example of the use of tie-beams is at Névache (Hautes-Alpes) built between 1490 and 1496. It is one of a number of churches on the eastern borders of France which continued to use timber reinforcement long after churches elsewhere had turned to iron. The exceptions to this generalization are the great Gothic cathedrals of France.

The earliest Gothic building to contain wooden reinforcement is the cathedral of Laon, built between 1160 and 1220 (Broche). The tie-beams were apparently inserted as a temporary measure, for they have all been sawn off flush with the stonework. They were used throughout the cathedral. In the choir, iron reinforcement, 2-in. (5-cm.) square, was anchored to iron hooks that still survive. Chartres Cathedral had the aisles of the nave and the choir reinforced with tie-beams. The voids remain where the sawn-off stubs have rotted away in the masonry. Reims Cathedral was similarly reinforced, but with iron as well as timber. The timber beams were placed throughout the aisles and across the ambulatory at the level of the springing of the arches of the lowest arcade. In the last few western bays of the nave aisles, iron hooks remain to show that tie-rods were used instead of tie-beams (pl.26). In these bays, there are hooks for arcade ties, but there is no arcade reinforcement elsewhere. There are iron hooks, too, set into the crossing piers and in the nave walls at the level of the base of the triforium chamber. The flying buttresses are underpinned by timber beams fixed between the buttresses and the cathedral walls. The east end of the cathedral served as a model for the choir of Beauvais which also contains tie-rods. Reims was built between 1212 and 1240 (Saint-Paul). Soissons Cathedral dates from between 1212 and c.1250 (Brunet). The nave, choir, transepts and ambulatory were reinforced with tie-beams and tie-rods. A restoration of the building undertaken by Brunet in the 1920s involved the rebuilding of a pier and resulted in the

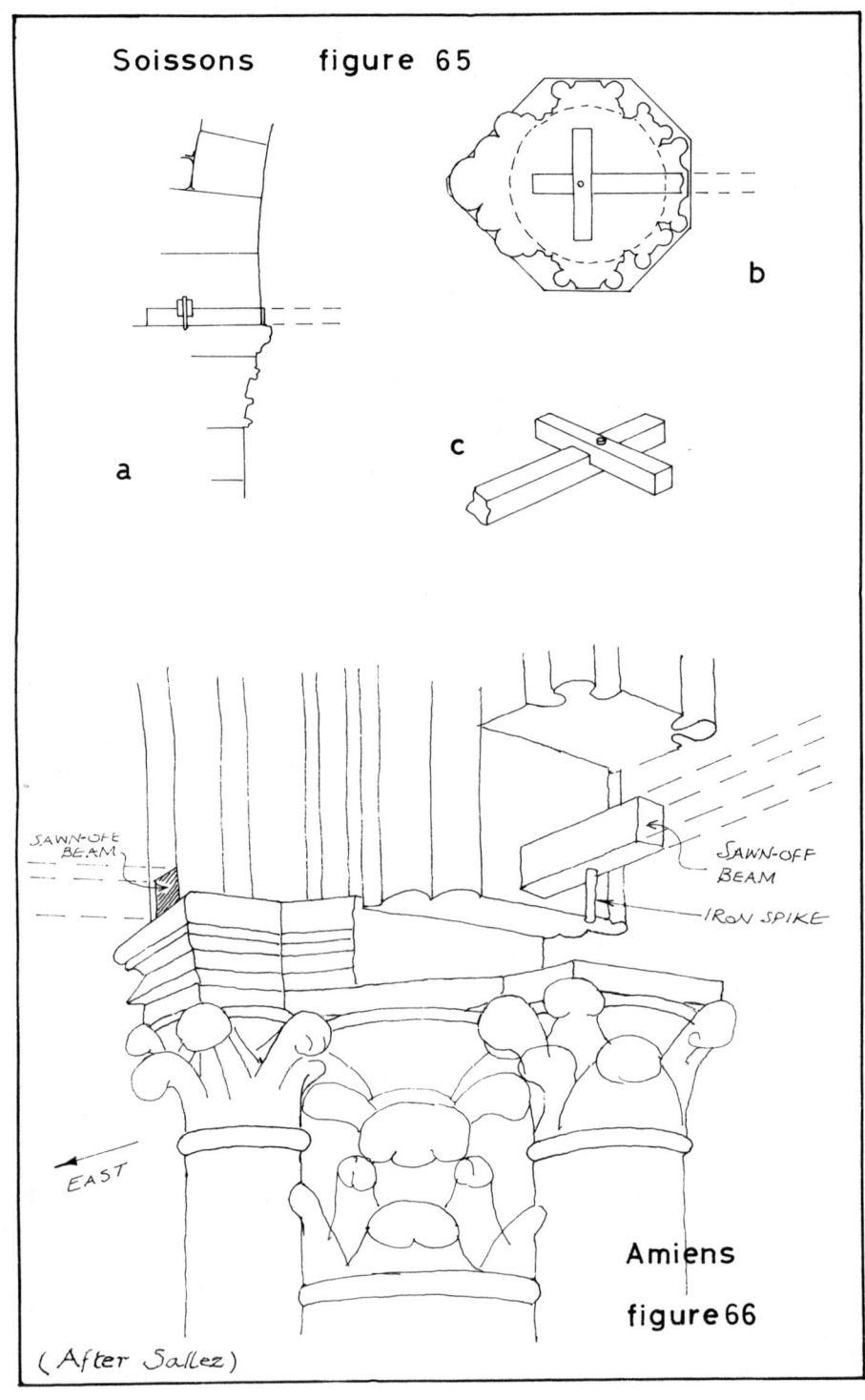

Soissons   figure 65

a

b

c

SAWN-OFF BEAM

SAWN-OFF BEAM

IRON SPIKE

EAST

Amiens
figure 66

(After Sallez)

discovery of the remains of part of a timber reinforcement embedded in the pier (fig. 65). It consisted of two beams jointed together at right-angles and further secured with an iron pin. Amiens Cathedral, built between 1220 and 1288 (Durand, 1901–3), was reinforced with timber ties across the nave[9], in the arcades and across the aisles throughout, except in the ambulatory, where iron hooks are still *in situ*. In some cases in the arcades, the timber voids are at some distance above the usual position at the springing of the arches. A restoration undertaken in the 1960s (Sallez) found the remains of the timber ties embedded in the piers (fig. 66). The ends of the tie beams were not jointed into a cross-piece as at Soissons, but each stopped short of the centre of the pier so that four separate pieces of wood were discovered in a pier. Each piece was fixed in position with an iron spike that was embedded into the masonry beneath the tie-beam. As at Reims, timber reinforcement was used between the flying buttresses and the walls of the cathedral.

An early example of metal reinforcement outside Burgundy appears in a church in as secluded an area as Limousin. This is in the village church of Chambon-sur-Vouieze (Creuse). It is a small building and was built during the early part of the twelfth century (*Limousin Roman*). In the second half of the century, iron reinforcement was used in the Cistercian abbey church at Noirlac (Cher) where the nave has a quadripartite vault supported by transverse arches supported by transverse arches and reinforced by iron tie-rods. The hooks for these still remain in the arcades of the nave not, as usual, at points in the springing of the transverse arches but above the arches of the nave arcade (pl.27). The church dates from between 1150 and 1170. At about the same time, the abbey of St. André-le-Bas at Vienne (Isère) was built (Aubert, 1965). Its church is rather like Autun Cathedral in style and had double tie-rods across the nave which was completed by 1152. Another church rather like Autun is St. Mammès, the cathedral of Langres (Haut-Marne). The nave vault is supported by pointed transverse ribs and was reinforced with iron tie-rods across the springing of these ribs. The iron fixings for these rods still remain (fig. 68).

From the area of Burgundy, the use of tie-rods was spreading out in two main directions: north-eastwards to the Rhineland and north-westwards towards Paris. The Rhineland will be referred to later, but in the other direction there is the collegiate church of St. Evremond, Creil (Oise), built *c.* 1160 (Gall, E., 1955). This is an early example of tie-rods used in a Gothic context and was closely followed by two churches in Paris. The cathedral of Paris was constructed between 1163 and 1182 (Cali) and still contains iron hooks in the ambulatory and in the piers of the choir that support the ribs of the vault. St. Julien-le-Pauvre, a Cluniac building, dates from between 1170 and 1220 and contains tie-rods across the nave and across the choir. The nave tie-rods are original while those in the choir are modern. Other churches of this period with reinforcement include St. Aignan at St. Aignan (Loir-et-Cher) with iron reinforcement in the crossing. In the same department is the church of Rhodon which dates from the thirteenth century (Leseuer) and has tie-rods across the nave. St. Quentin (Aisne) has a choir built in the early thirteenth century with tie-rods which were further reinforced with flying buttresses in 1316. Tie-rods were later introduced across the nave between 1468 and 1475. In the department of Bas-Rhin is Ste.-Foy at Sélestat with a ribbed vault with transverse arches. Each alternate pier is reinforced with tie-rods across the

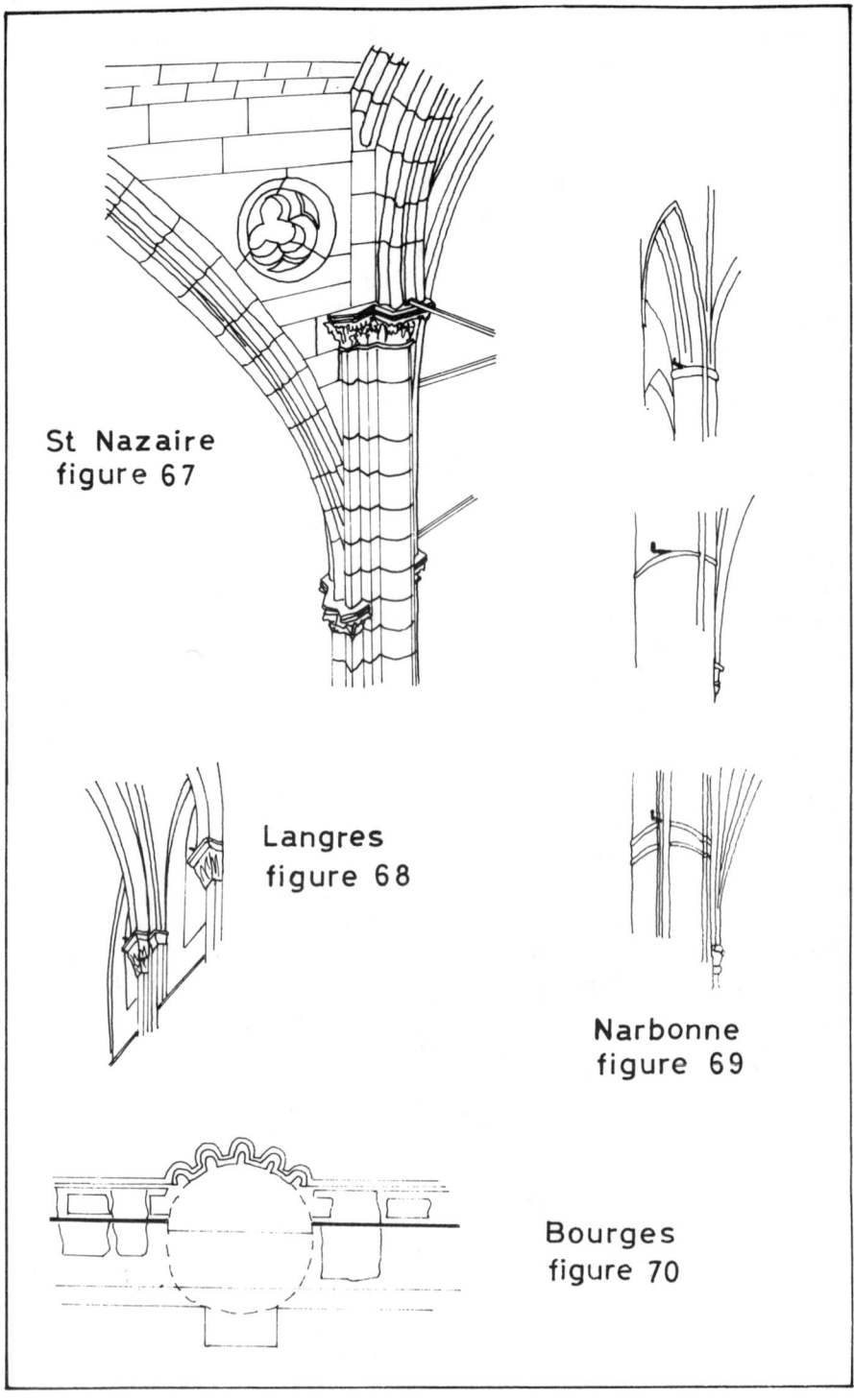

St Nazaire
figure 67

Langres
figure 68

Narbonne
figure 69

Bourges
figure 70

nave. The church dates from the later twelfth century. Rosheim, a little later, has tie-rods across the aisles.

The cathedral of Bourges (1230–1330) has the floor of the triforium chamber strengthened with continuous iron chains shown in plan in fig. 70. Two Cluniac churches of the same period are the abbey at Longport (Aisne) and St. Leu-d'Esserent (Oise). The first was built in the style of Soissons and today it lies in ruins, but it is possible to see the voids in the nave arcades where ties across the aisles were once placed. In a few places, the tie-rods are still in position. The church was built between 1200 and 1227. St. Leu-d'Esserent dates from 1230 and has tie-rods across the nave. Tournai Cathedral contains tie-rods in the choir which was built between the years 1242 and 1325.

Two churches close together in the south-west are Narbonne Cathedral (Aude) and St. Nazaire, Carcassone. Narbonne, like Tours, has a choir modelled on that of Amiens, and is built throughout in the northern Gothic style with choir arcades that were once reinforced with iron tie-rods (fig. 69). The church dates from between 1273 and 1340 (Christ). St. Nazaire was also influenced by northern examples and the choir was completed by about 1300 (Van der Meer) and is tied with iron rods (fig. 67).

In Burgundy, the cathedral of Auxerre was built between 1277 and 1300 (Porec) with part of the east end reinforced with tie-rods. Not far off is the church of Mussy-sur-Seine begun in 1300 (Branner). Iron tie-rods reinforce the nave. Beauvais Cathedral was built between 1284 and 1324 after the vaults of the previous building had fallen (Cali). The choir was inspired by Amiens and was reinforced by tie-rods in the arcades and across the ambulatory. The rods have now gone but the iron hooks remain. The flying buttresses of the chevet, however, still have tie-rods. To the east of Paris, in the department of Marne, the church of Notre Dame de l'Epine was built between 1410 and 1470 (Luc-Benoist). The nave and transepts date from after 1439 and were reinforced with iron. The church is a copy of Reims Cathedral and the surviving iron hooks are identical with those on the western bays of the nave at Reims.

This concludes the survey of the churches of France that used reinforcement. We have dealt with the spread of the technique from the Burgundian 'heartland'. This spread carried with it the use of both materials, timber and iron, and we find, in the case of the great Gothic cathedrals, that both materials were commonly used in the same church. In these churches the actual ties have usually been removed, and so regularly removed in the case of the timber ties, that it is clear that their function was viewed as a purely temporary one, viable only until the church had achieved a permanent stability. This may have been a period of many years, for the mortar used at that time hardened at a very leisurely rate. This was especially true of the lime mortars native to France which set very slowly when unexposed to the air. As Fitchen (note 193, p. 226) observes:

> 'It would seem that most of the dimensional discrepancies and departures from geometrical accuracy in the existing structures of almost every medieval church in France are due in large part ot the nature of the mortar and its excessively slow setting time'.

One is tempted to believe that reinforcement was used to counteract this disadvantage inherent in the mortar. Certainly, the extensive use of timber in a cathedral like Amiens would have represented a large investment in terms of money and labour which would not have been contemplated unless it performed a very real function in the construction of the building. The prevention of distortion during the settling of the building is the obvious function to suggest. In buildings outside the Ile de France it was more usual to leave the reinforcement *in situ*, and it is difficult to suggest why this should be so. It may be that it was only in the greater churches that reinforcement was thought to intrude on the ensemble; in lesser churches it might have been ignored as a necessary evil. Perhaps a difference in taste between metropolitan and provincial areas was the main factor.

Elsewhere in France, apart from Burgundy and the Ile de France, the use of reinforcement was not very common. The number of examples known to the writer are as follows: in Provence, four; in the Auvergne there are no examples; in Languedoc, two; in Poitou and Perigord, three; in Normandy there are no examples; in the centre of France (Loir-et-Cher, Cher and Creuse) there are four examples, and in Brittany there is one. These figures are to be compared with sixteen in Burgundy and sixteen in the Ile de France. The chronological progression in the area now known as France is undoubtedly Burgundy and then the Ile de France with the examples outside these areas probably influenced by Burgundy during the Romanesque period and by the Ile de France during the Gothic period.

The other important area that can be described as an extension of Burgundy is the Rhineland. The Byzantine influence at St. Maria im Kapitol in Cologne has already been suggested. This church has no close analogies elsewhere in Germany and there is a gap of a hundred years before reinforcement appears in the Rhineland again. When it does so, it is timber reinforcement at the Benedictine abbey of Maria Laach, 20 kilometres west of Coblenz. The church is very largely a copy of Vézelay, with the same architectural vocabulary: groin vaulting, transverse arches, absence of triforium and reinforcement. Before the Second World War there were tie-rods across the nave between the springers of the transverse arches, but they were secured at the haunches of the arches instead of at the lower level of the springing as at Vézelay. Maria Laach is the first fully-vaulted basilica in the Rhineland and was built between 1130 and 1156 (Gall, E., 1963). The nave at Vézelay in its original form was built between 1120 and 1132, so that the copy at Maria Laach lacks the flying buttresses that were later added to Vézelay when the tie-rods were found to be ineffective. Obviously the reinforcement in the German church was more effective, probably because its operation took place higher up the haunches of the vaults than at Vézelay. Another church using the same architectural vocabulary and which is also a copy of the Burgundian church is the abbey at Knechtsteden near Anstel. The iron tie-rods are secured in the same way as those at Maria Laach which was under construction at the same time as this church (Gall). A similar architectural tradition, somewhat modernised, was continued in the Benedictine abbey of Brauweiler which has a ribbed vault constructed after 1141 (Gall). This church has both a decorated tie-beam and tie-rods which are probably later replacements of the original rods. Strands of the Romanesque tradition established at Maria Laach can be seen in the cathedral at Worms built between 1181 and 1234 (Conant). There are ribbed vaults with pointed arches, iron tie-rods across the nave and no flying buttresses.

The church of the Holy Apostles, Cologne, has a trefoiled plan like the church at St. Maria im Kapitol in the same city, but the reinforcement owes nothing in derivation to the earlier church. Instead it is derived from the Burgundian/ Rhenish practice and consists of tie-rods across the nave. The present tie-rods are modern and replace originals destroyed during the Second World War like much of the reinforcement in the Rhineland. The church dates from between 1192 and 1230 (de Lasteyrie). Another Benedictine abbey is the Minster at München-Gladbach, formerly St. Vitus. It has a pointed quadripartite rib-vault reinforced, between the springings of the transverse arches, with metal tie-rods which are almost certainly replacements of earlier ties. The building was put up between 1200 and 1242 (Gall). The Minster of St. Quirinus at Neuss, begun in the year 1209 (Gall) has a ribbed vault with transverse arches which are reinforced with tie-rods (pl.28). A church with a similar ensemble is at Boppard about twelve kilometres from Coblenz, dating from 1230 (Busch). Tie-rods cross the nave. At Oberwesel, the collegiate church of Our Lady has a pointed rib-vault with iron tie-rods reinforcing the choir and apse and was built during the first half of the fourteenth century (Gall). Finally, we have a Gothic example of iron reinforcement. This is the Heiliggeistkirche in Heidelberg which has tie-rods in the nave arcades and dates from the fifteenth century (Stich).

The Rhineland, speaking solely in terms of reinforcement, does seem to have kept fairly well to the tradition established by Maria Laach and derived from Burgundy. There is only one instance of timber reinforcement in this series and this is used in combination with tie-rods at Brauweiler. Apart from that, tie-rods were employed across the naves between the springings of the transverse arches. The only churches that do not conform to this pattern are the two latest examples, the churches at Oberwesel and at Heidelberg.

Further south, in Austria, there are two churches containing iron reinforcement. One is the Stiftskirche at Seckau, built between 1142 and 1164 (Puhringer) with a rib vault. It has tie-rods across the nave. The other is the cathedral at Gurk, erected between the years 1164 and 1174 (Puhringer) with a similar rib vault reinforced in the same way with tie-rods. These two churches are on the route from northern Italy into Austria and one should perhaps associate them with Yugoslavian churches and ascribe their metal reinforcement to influences emanating from northern Italy. These Yugoslavian churches include Korcula Cathedral erected during the thirteenth century with tie-rods, Trogir Cathedral with tie-rods fixed across the nave and through the clerestory windows (fig. 49 and pl.12) dating between 1206 and 1255 and the cathedral at Sibenik (*c.* 1400) with tie-rods throughout.

Also east of the Rhineland is the town of Regensburg which contains the church of St. Jakob, built in the later twelfth century. The reinforcement in this building was of timber; the holes into which the cross-beams were fixed still remain each side of the nave.

An interesting development of the Venetian use of tie-beams is their transmission to northern Germany and Poland in the wake of the Teutonic Knights. The historical background starts soon after the Grand Master of the Teutonic Order had diverted the energies of the Knights from the Holy Land to Hungary. An agreement between them and the Polish Prince Conrad of Mazovia switched the attention of the Order to northern Europe. A crusade began against the pagans of the area. Between 1228 and 1230 they began the conquest of Prussia and, with the help of both Pope and Emperor, they completed it within fifty years. They introduced large numbers of German settlers, forcibly converted the peasants to Christianity and built castles and churches. After 1283, they annexed Danzig and Pomerellen and, in 1309, the Grand Master transferred his official residence from Venice to Marienburg in Prussia and the order set up as a territorial German power under the nominal suzerainty of Pope and Emperor. The Order ruled the coastands from Livonia to Pomerellen and, later, Danish Estonia on the Gulf of Finland. Influences emanating from the headquarters of the Order in the Italian city could have been responsible for the Venetian characteristics of brick construction (eminently suitable for the stoneless Baltic littoral) and the tie-beams that were sometimes decorated. Timber, of course, was abundant in the area.

These architectural traits first appear at Altenkrempe (Holstein) at the western end of what was to become the Teutonic coastland. The church, built there between 1190 and 1244 (Busch) contains decorated tie-beams across the nave. The Knights quickly spread along the Baltic shore and were swiftly followed by commercial interests that resulted in the formation of the Hanseatic League of Hamburg and Lubeck in 1241. These two towns were later joined by others further east; Stralsund and Danzig were the most important of these. Lubeck became the capital of the League. After 1251, the church of St. Mary (Clasen) was built there and it became the model for several great brick churches of the region. St. Mary is reinforced with tie-beams across the nave. At the other end of the Teutonic littoral, the city of Danzig felt the influence strongly enough before its annexation by the Teutonic Knights to build a Marienkirche with tie-beams and prop-beams in the choir and the choir aisles. The church dates from between 1220 and 1266 (Drost).

Greifswald became one of the Hanseatic towns. It has a Marienkirche, built as a hall church, with tie-beams across the nave (fig. 72) in the second half of the thirteenth century (Clasen). The prospering burghers in Stralsund completed their Marienkirche in 1298 (Clasen) reinforced with tie-beams across the nave and in the nave arcades (fig. 71). A little later they repeated the venture, completing a Nicolaikirche in the same town in 1311 (Clasen). It has tie-beams placed across the nave. In Wismar, another Hanseatic town, the church of St. Marien was begun in 1339 (Gross), but the present tie-beams, decorated and high up, spanning the nave, date from much later and probably replaced the original tie-beams. In the mid fourteenth century, the church of St. George was completed in the same town with tie-beams across the nave. A third church in this town, begun in the late fourteenth century but not completed until 1459 (Clasen), was the Nicolaikirche with timber reinforcement in the same position. A last example in this region is the only monastic church in the list. It is the Klosterkirche, a Cistercian foundation built in another Hanseatic town, the city of Doberan. Both nave and nave arcades are reinforced with timber beams.

Britain was on the periphery of the area where reinforcement was commonly used. It was not part of the tradition of masonry church building in Anglo-Saxon times as far as we know. Churches in Normandy did not contain it; the self-confident Norman builders felt no need for it and, in Britain, no Norman churches contain examples. The only area where Norman building is associated with reinforcement is southern Italy and Sicily where Byzantine structural techniques were an important factor in local building styles.

In Britain it is only when we come to Gothic churches which were influenced to

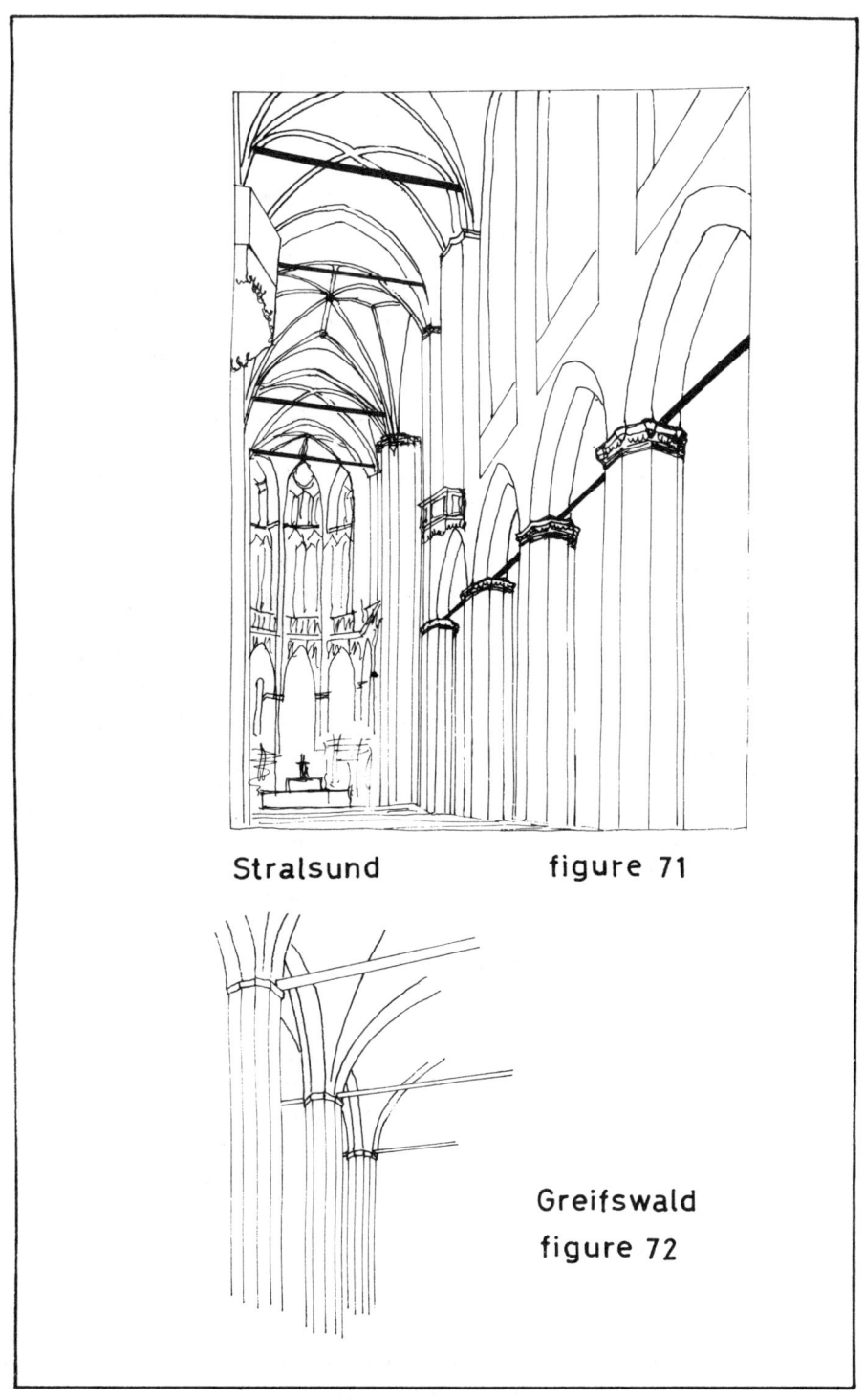

Stralsund   figure 71

Greifswald
figure 72

some extent by buildings in the Ile de France that we can find examples of reinforcement. However, they are few, and the idea did not commend itself to English builders who were not interested in erecting vaults of great height where the problems of achieving stability are much more difficult.

The earliest example of reinforcement in English cathedrals is at Canterbury where the choir was begun by William of Sens. Sens Cathedral nave, begun *c.*1140, has iron reinforcement in the arcades, and it may be that this is an example of a direct constructional influence passing via the builder from one country to another. Canterbury choir was begun in 1176 but, before it was completed, the French mason fell from the scaffolding and was so badly injured that he had to relinquish the contract. An English master mason took his place but it can be seen how the work continued in the same style. Iron tie-rods were dutifully inserted and were perhaps made necessary because of the French design. The eastern limb of Canterbury was completed in 1184 (Harvey, J.). At Lincoln there are two wooden prop-beams across the entrances to the eastern transepts. The earlier apsidal presbytery, its processional path and chapels were pulled down and five new bays substituted. As a precaution the prop-beams were put in at the western end of these new bays but were never removed and have become a permanent feature of the cathedral. The prop-beams were inserted after 1255.

The best example of reinforcement in an English church is at Westminster Abbey. The eastern end is strengthened both with wooden and iron reinforcement and was built by Henry of Reyns between 1243 and 1253 (Harvey, J.). This is perhaps one of the constructional details in Westminster Abbey which is due to French influence. The original iron strapping in the tower and spire of Salisbury cathedral was described by Wren in 1668 as very fine and at his suggestion extra iron bands were inserted. The originals dated from the period when the spire and upper part of the tower were built, beginning in the 1330s.[10]

The last part of this article has dealt with the techniques of timber and iron reinforcement in Eastern and European church building from the fourth century to the fifteenth. During this period the techniques spread from their place of origin in the countries bordering the eastern Mediterranean to much of Europe and, in the thirteenth century, crossed the English Channel to appear briefly in Britain at the furthest end of the trans-continental axis.

This expansion, in so far as it can be said to have had a single direction, passed through several regions important in the history of European architecture: Byzantium, Lombardy, Burgundy, northern France and the Rhineland. All but the last of these regions seems to have incorporated the techniques into strands of its own tradition and then passed them on to neighbouring areas. In this fashion, the techniques contributed something at most stages to the development of European church architecture and played a part in its fruition in northern French gothic.

Acknowledgments

I should like to acknowledge first of all the assistance of Dr. John Harvey to whose wide scholarship are owed many suggestions and emendations. I am only sorry not to have had the space to develop more of his suggestions. Thanks are due to Mr.

John Fitchen and his publishers, the Oxford University Press, and to M. A. Sallez for permission to reproduce some of their published material. Others to whom I am indebted are Mr. Peter Addyman, Mr. B. J. Ashwell (Architect to the Dean and Chapter of Gloucester Cathedral), Professor Michel de Bouard, Professor Kenneth J. Conant, Professor Sumner McKnight Crosby, M. B. Darré (Centre de Recherches des Monuments Historiques), Professor Martyn Jope, Dr. Peter Kidson (Courtauld Institute), Mr. David King, the Librarian and Staff of the Library of the Royal Institute of British Architects, Mr. A. T. Lucas (National Museum of Ireland), Mr. Herman G. Ramm (Royal Commission on Historic Monuments), Mrs. D. Roddham, the late Mr. L. Salzman, Mr. R. O. Spring (Clerk of Works, Salisbury Cathedral), Dr. M. W. Thompson (Inspectorate of Ancient Monuments, Department of the Environment) and Mr. Richard Warmington. This list could be continued to include many people from whose knowledge information has been drawn, both by way of the printed word and illustrations and by word of mouth on my travels. These latter include a village priest who got out of bed and cycled two miles with his cassock over his pyjamas to show me round his parish church, which he did in Latin. I was only able to understand a shameful fraction of his discourse but what did come across strongly was his enthusiasm and love for the building. This is one of the most heartening things about the study of ecclesiastical architecture—the pride that people take in their local monuments. To them and especially to the good father 'Gratias maximas ago vobis'.

NOTES

[1] The following passage has been translated by Michael Avery from du Pontet's O.C.T. edition of Caesar's *De Bello Gallico*, 7.23.

All Gallic walls are built more or less as follows. Straight beams are placed on the ground throughout the length of the wall at equal intervals, with a two foot gap between each one and its neighbour. They are fixed together inside the wall and covered with a quantity of bank material, and the gaps I have mentioned are blocked up on the front face with sizeable rocks. When these are in position and packed tight, another layer is added on top, maintaining the gap already mentioned and with the beams not touching one another but placed with equal gaps between each pair, and each beam skilfully held in position by the stones placed between them. The whole construction is put together until it reaches the correct height for the wall. Not only is the variegated appearance given by this construction attractive to look at, with alternate ranks of timbers and stones each in a quite straight line, but it is extremely suitable for use for town defences, since it is protected from fire by the stone and from battering rams by the timbers, which, consisting usually of continuous forty-foot lengths fixed together inside the wall, cannot be smashed through nor torn apart.

If one attempts to reconstruct a Gallic wall from Caesar's description, it is immediately apparent how vague it is. All that one can adduce from it in strictly archaeological terms is that the walls were built of stones, 'bank material' (earth?) and beams probably fixed together with nails. The important sentence is the first one, which probably indicates not only that this is a composite description with features drawn from several sources, but also that there was no standard method of wall building among the Gauls, and that the materials were ubiquitous while the design varied from site to site.

[2] The following passage has been translated by Michael Avery from Granger's Loeb text of Vitruvius's *De Architectura*, 1.5. paragraph 3.

The thickness of a city wall, I feel, should be enough for two armed men moving in opposite directions along the top of it to pass when they meet without getting in one another's way.

Lengths of charred olive wood should be put in position as close to one another as possible throughout the length of the wall, so that the two faces of the wall are fastened together by these timbers, as if by safety pins, to give permanent strength; for neither rot, weather nor ageing damages this kind of timber, which remains serviceable even if buried underground or placed in water. Not only a city wall but also substructures and any internal walls which need to be made as thick as a city wall will last undamaged for ages if they have ties in this way.

Vitruvius emphasizes a point not made by Caesar: that the two faces of the wall should be held together by transverse timbers, while he makes no mention of the longitudinal beams that are the main feature of Caesar's account. Vitruvius's description brings to mind the use of timber at Knossos where transverse beams were used, also in an 'architectural' environment.

[3] P. V. Addyman. Report forthcoming.

[4] Personal communication from the excavator, Herman G. Ramm.

[5] I am grateful to Mr. James Northeast for pointing out this example to me.

[6] I owe this information to Dr. T. McNeil.

[7] 'A quavery or maris and unstable foundacion must holpe with great pylys of alder, rammed down with a frame of tymbre called a crossaundre'. Horman. *Vulgaria*. C. XXIX.

[8] I owe much of the information and the drawings in this section to Mr. John Fitchen's *The Construction of Gothic Cathedrals* (Oxford, 1961).

[9] Another cathedral that was timber-reinforced throughout like Amiens is Tours. The choir was modelled on Amiens and dates from the mid thirteenth century: the nave was built between *c*.1260 and 1266 (Christ).

[10] I owe this information to the Clerk of the Works at Salisbury, Mr. R. O. C. Spring.

BIBLIOGRAPHY AND REFERENCES

Agnello, G. *L'Architettura bizantina in Sicilia* (Florence, 1952).

American Institute of Architects' Journal (New York, Oct., 1965).

Andrews, F. B. *The Medieval Mason and his Methods* (Oxford, 1925).

Afray, M. *L'Architecture religieuse du Nivernais au moyen age—les églises romanes* (Paris, 1951).

Allen, J. *Cornish Archaeology*, vi (1967).

Anonymous: 'The Coptic Churches of the Wady-el-Natrun, Lower Egypt'. *The Builder*, Oct. 13, 1911.

Archaeologia Cantiana, xlvi, 179–95.

'Archaeological Intelligence', *Archaeological Journal*, iii (1848), 36.

L'Architettura Cronache e Storia. ii (1956/7), v (1959/60), 560–5, viii (1962/3), 194–201, x (1964/5), 194–201, xi (1965/6), 50–7, xiv (1968/9), 554–57.

Arslan, E. *Vicenza. I. Le chiese* (Rome, 1956), pp. 156–61.

Arslan, W. *L'Architettura romanica veronese* (Verona, 1939), pp. 20–4, 87–90.

Aubert, M. *L'Architecture cistercienne en France* (Paris, 1947).

Aubert, M. *Romanesque Cathedrals and Abbeys of France* (London, 1966), pp. 153–5, 254–9, 257–8, 262–3, 376–7, 449.

Bachman, W. *Kirchen und Moschen in Armenien und Kurdistan* (Leipzig, 1913).

Badaway, A. *Architecture in Ancient Europe and the Near East* (1966), pp. 60, 120–1.

Ballance, S. 'Nineteenth century monuments in the City and Vilayet of Trebizond'. APXEION ΠONTOY (1966).

Bandmann, G. *Mittelalterliche Architektur als Bedeutungstragers* (Berlin, 1951).

Barbacci, A. *Il restauro dei monumenti in Italia* (Rome, 1956).

Baselli, V. Monograph in *Rivista Archaeologica della Provincia di Como*. Fasc. 30. 2 vols. (October 1887 and August 1872).

Bausson, C. *Vieilles églises de campagne* (Paris, 1964).

Baum, J. *Romanische Baukunst in Frankreich* (Stuttgart, 1910), pp. 84–5.

Bell, E. *Pre-Hellenic Architecture in the Aegean* (1926), p. 213.

Bernady, A. A. *L'Istria e la Dalmazia* (Bergamo, no date).

Bersu, G. *Das Wittnauer Horn* (1945) and *Antiquity*, xx (1946), 4–8.

Bertaux, E. *L'Art dans l'Italie méridionale*. 2 vols (Rome, 1904).

Beyli, L. de *L'Habitation byzantine* (Grenoble and Paris, 1902), p. 83.

Bianchi, L. *La pieve di S. Michele Arcangelo in Nonantola* (Rome, 1937), p. 37.

Biddle, M. *Antiquaries Journal*, xlvii (1967), 268–72.

Blegen, C. W. (editor). *Troy*. Vol. iii (Cincinatti, 1935).

Bocazzi, F. Z. *La basilica dei Santi Giovanni e Paolo in Venezia* (Venice, 1966).

Bock, W. de *Matériaux pour servir l'archéologie de l'Egypte chrétienne* (St Petersburg, 1901, pp. 88–90.

Boskovic, D. *Medieval Art in Serbia and Macedonia* (Belgrade, 1948), p. 54.

Branner, R. *Burgundian Gothic Architecture* (London, 1960), pp. 99, 139–40, 186–7.

Branner, R. *La Cathédrale de Bourges et sa place dans l'architecture gothique* (Paris and Bourges, 1962), p. 83.

Broche, L. *La Cathédrale de Laon* (Petites Monographies des Grands Edifices de la France, Paris, 1926).

Bross, J. (editor). *Histoire générale des églises de France, Belgique, Luxembourg et Suisse* (Paris, 1966).

Browne, E. A. *Early Christian and Byzantine Architecture* (London, 1912).

Brunet, E. 'La restauration de la Cathédrale de Soissons'. *Bulletin Monumental*, lxxxvii (1928).

Bulletin Monumental, lxvii (1903): article on Chartres; lxviii (1904): article on St. Trophime.

Brusin, G. *Aquileia. Padua* (1969), p. 13.

Bumpus, T. F. *The Cathedrals and Churches of Northern Italy* (London, 1907), p. 34.

Bumpus, T. F. *The Cathedrals and Churches of the Rhine and North Germany* (London, 1946), p. 76, fig. on p. 77.

Burstow, G. F. and Honeyman, G. A. *Sussex Archaeological Collections*, cii (1964), 55–67.

Busch, H. *Germanica Romanica* (Vienna and Munich, 1963), pp. 272, 289ff., 307.

Bushe-Fox, J. P. *Journal of Roman Studies*, xxii (1932), 60–6.

Butler, A. J. *Ancient Coptic Churches of Egypt* (Oxford, 1884), pp. 18, 135–48, 150.

Butler, H. C. *Architecture and other Arts*. Part II of the *Publications of an American Archaeological Expedition to Syria in 1899–1900* (London and New York, 1904).

Butler, H. C. *Early Churches in Syria* (Princeton, 1929).

Cadafach, Puig i. *L'arquitectúra romànica a Catalunya*. 2 vols (Madrid, 1911).

Cali, F. *L'Ordre ogival* (Paris, 1963), pp. 251–2.

Canale, G. G. *Strutture architettoniche normane in Sicilia* (Palermo, 1959).

Caroe, A. D. R. *Old Churches and Modern Craftsmanship* (Oxford, 1949).

Carswell, J. *New Julfa. The Armenian Churches and other Buildings* (Oxford, 1968).

Cattaneo, R. *Architecture in Italy from the Sixth to the Eleventh Centuries*. Translated by Countess Isabel Curtis (London, 1896).

Caumont, Arcisse de. *Abécédaire d'archéologie. Architecture civile et militaire* (1870), pp. 359, 513–14.

Ceschi, C. *Architettura romanica genovese* (Milan, 1954), pp. 31–43, 110–17, 205–9.

Choisy, A. *L'Art de bâtir chez les Byzantines* (Paris, 1883), p. 9, fig. 4.

Choisy, A. *Histoire de l'architecture*. 2 vols (Paris, 1899), pp. 59, 160–2, 181, pl. i.

Christ, Y. *Cathédrales de France* (Paris, 1963), pp. 84, 138, 163–6.

Clapham, A. W. *Antiquaries Journal*, viii (1928), 350–3.

Clark, G. T. *Medieval Military Architecture*, i (1884), pp. 284, 451, 466, 481, 485, 501.

Clarke, S. *The Christian Antiquities in the Nile Valley* (Oxford, 1912).

Clasen, K.-H. *Die gotische Baukunst* (Potsdam, 1930), pp. 121, 129, 158, 163–4.

Clifford Perks, J. *Proceedings of the Dorset Natural History and Archaeological Society*, lxxvi (1954).

Conant, K. J. *Carolingian and Romanesque Architecture, 800–1200* (London, 1959), pp. 81–2, 129–30, 230, 241, 244, 248, 250, 251, 262.

Conteneau, G. *La Civilisation des Hittites* (Paris, 1948).

Corkhill, T. *A Glossary of Wood* (London, 1948).

Corroyer, E. *Gothic Architecture*. Translated by W. Armstrong (London, 1813).

Cotton, M. A. 'British Camps with Timber-laced Ramparts'. *Archaeological Journal*, cxi (1955), 26–105.

Country Life (1914), pp. 422–27.

Creswell, K. A. C. *Early Moslem Architecture*. 2 vols (Oxford, 1940), i, p. 404.

Creswell, K. A. C. 'Fortification in Islam before A.D. 1250'. *Proceedings of the British Academy*, xxxviii (1952), 114.

Crosby, S. McK. *The Abbey of St. Denis* (Yale, 1961).

Cumberland and Westmorland Archaeological Society Transactions, xxvii (1927), 224–6.

Cunliffe, B. *Antiquaries Journal*, xliii (1963), 218–27.

Cunliffe, B. *Iron Age Communities in Britain* (London, 1974).

Curwen, E. C. *Antiquaries Journal*, xii (1932), 1–16.

Dartein, F. de. *Etude sur l'architecture lombarde et sur les origines de l'architecture romano–byzantine* (Paris, 1865–82).

Davies, J. G. *The Origin and Development of Early Christian Architecture* (London, 1952).

Debret, F. 'Notice sur les diverses constructions et restaurations de l'Eglise St-Denis'. *Séance publique des cinq académies* (Paris, 1842), pp. 9–28.

Dechelette, J. *Manuel d'archéologie*, v (1931), pp. 530–1.

Decker, H. *Romanesque Art in Italy* (London, 1958).

Dereko, A. *Monumental and Decorative Architecture in Medieval Serbia* (Belgrade, 1953), pp. 273, 336–7, 349, pls. 217, 253, 259, 262, 336, 350, 352, 355.

Der Nersessian, S. *Armenia and the Byzantine Empire* (Cambridge, Mass., 1947).

Dictionary of the Architectural Publication Society, i (1856).

Dictionnaire des églises de France, iv (Paris, 1968), iv. A. 119 and iv. C. 83.

Diehl, C. *Manuel d'art byzantin* (Paris, 1910), pp. 138–9.

Diehl, C. *Les monuments chrétiens de Salonique* (Paris, 1918), pp. 189–202, 218–9, fig. 96, pls. LII, LVI, LXI ser.

DoE: Department of the Environment.

Drack, W. *L'Age du Bronze en Suisse* (1959).

Drost, W. *Die Marienkirche in Danzig und ihre Kunstschatze* (Stuttgart, 1963).

Durand, G. *Monographie de l'église Notre-Dame, cathédrale d'Amiens* (1901–3).

Durand, G. *Eglises romanes des Vosges* (Paris, 1913).

Ebersolt, J. *Orient et occident* (Paris and Brussels, 1928).

Ebersolt, J. *Monuments d'architecture byzantine* (Paris, 1934), pp. 82, 88, 147.

Enlart, C. *L'Art gothique et la renaissance en Chypre*. 2 vols. (Paris, 1899).

Enlart, C. *Manuel d'archéologie française depuis les temps merovingiens jusqu'á la renaissance*. 2 vols. Part I. *Architecture religieuse* (Paris, 1927), pp. 462, 573–4, Part II, p. 865, fig. 431.

Evans, Sir A. *The Palace of Minos*, vol. i (London, 1921).

Fitchen, J. *The Construction of Gothic Cathedrals* (Oxford, 1961).

Fleury, G. *La Cathédrale du Mans* (Petites Monographies des Grands Edifices de la France, Paris, 1910).

Flipo, V. *La Cathédrale de Dijon* (Petites Monographies des Grands Edifices de la France, Paris, 1928).

Focillion, H. *The Art of the West in the Middle Ages*. 2 vols (London, 1969).

Forchheimer, P. and Strzygowski, S. *Die Wasserbehälter von Konstantinopel* (Vienna, 1893), pp. 57ff., 63, 67.

Forlati, F. *Saint Sophia of Ochrida* (UNESCO, 1953), p. 9.

Forminge, J. *Saint-Denis*. Les Monuments Historiques de la France, No. 3 (July–September, 1955).

Frankl, P. *Die Frühmittelalterliche und romanische Baukunst* (Wildpark-Potsdam, 1926).

Franklin, J. W. *The Cathedrals of Italy* (London, 1958), pp. 128, 138–40, 158–67, 182–92, 266.

Frei, B. 'Die Grabung auf dem Montlingerberg' in *Jahrb. Schweiz. Gesell. Urgesch.* xliv (1954–5), 146–51.

Gall, E. *Die Gotische in Frankreich und Deutchsland*. Vol. i. *Nordfrankreich* (Braunschweig, 1955), pp. 202–3, 292–3.

Gall, E. *Cathedrals and Abbey Churches of the Rhine* (London, 1963), pp. 33–5, 50, 58, 65, 218–9.

Gall, L. *L'Architecture religieuse en Hongrie du XIe au XIIIe siecles* (Paris, 1929).

Gantner, J. and Pobé, M. *Romanesque Art in France* (London, 1956).

Gavini, C. I. *Storia dell'architettura in Abruzzi*. 2 vols (Milan-Rome, 1928), i, pp. 27–36, 90–92, 209–11; ii, pp. 5, 215, 392–5.

Gayet, Al. *L'Art copte* (Paris, 1902), pp. 158–60.

George, W. S. *The Church of St. Eirene at Constantinople* (Oxford, 1912), p. 11.

Gieure, M. *Les églises romanes en France* (Paris, 1954).

Gimpel, J. *The Cathedral Builders*. Translated by C. F. Barnes, Jnr. (London, 1961).

Godfrey, W. H. *Guide to Lewes Castle* (Sussex Archaeological Society, 1968).

Golzio, V. *Architettura bizantina e romanica* (Bologna, 1939).

Gómez-Moreno, M. *El arte románico espanol* (Madrid, 1934).

Grodecki, L. *L'Architecture ottonienne* (Paris, 1958).

Gross, W. *Die abendländische Architektur um 1300* (Stuttgart, 1948), pp. 52, 55, 124, pl. 76.

Gsell, S. *Les monuments antiques de l'Algérie*. 2 vols (Paris, 1901).

Guyer, S. *Siena and the Hill-towns of Southern Tuscany* (No publication source, no date).

Hamilton, J.A. *Byzantine Architecture* (London, 1945).

Harvey, J. *The Gothic World 1100–1600* (London, 1950).

Harvey, J. *The Cathedrals of Spain* (London, 1957), pp. 50, 106–7.

Harvey, J. *English Medieval Architects* (London, 1954), p. 275.

Harvey, W. *Structural Survey of the Church of the Nativity, Bethlehem* (Oxford and London, 1935), pp. 9–11.

Heer, F. *The Medieval World* (London, 1963).

Hoddinott, R. *Early Byzantine Churches in Macedonia and Southern Serbia* (London, 1963), pp. 155–8.

Hope, St. John, W. H. *Archaeological Journal*, xli (1884), 1–34.

Honour, H. *The Companion Guide to Venice* (London, 1965).

Houvet, E. *Monographie de la Cathédrale de Chartres. Architecture* (Paris, Academie des Beaux-Arts, 1925), pp. 7–8.

Howgrave-Graham, R. P. *The Cathedrals of France* (London, 1959).

Hubert, J. *L'Art pré-romain* (Paris, 1938).

Hubert, J. *L'Architecture religieuse du haut moyen âge en France* (Paris, 1852).

Hunter Archaeological Society, iv (1927), 7–27.

Innocent, C. F. *The Development of English Building Construction* (1916), p. 101.

Jackson, T. G. *Dalmatia, the Quarnero and Istria*. 2 vols. (Oxford, 1870), i, pp. 136–7, 368; ii, p. 237.

Jantzen, H. *Ottonische Kunst*. 3 vols. (Munich, 1947), pp. 33, 41, 43.

Jones, H. J. E. *The Monasteries of the Wady-el-Natrun*. Bulletin of the Museum of New York, vi (1911) and vii (1912).

Jope, E. M. (editor). *An Archaeological Survey of County Down* (HMSO, 1966), pp. 290–1, 306–7, 409–11.

Jope, E. M. *The Leaning Tower of Bridgnorth. Timber lacing in the Twelfth Century* (unpublished).

Jorga, N. *Histoire de l'art roumain ancien* (Paris, 1922).

Kautzsch, R. *Romanische Kirchen in Elsass* (Freiburg in Breslau, 1927).

Kautzsch, R. *Kapitellstudien* (Berlin, 1936), p. 163.

Kenyon, K. *Digging up Jericho* (London, 1957).

Kirchmayer, M. *L'Architettura italiana dalle origini al giorni nostri*. 2 vols. (Turin, 1958).

Koldeway, R. *Das wiedererstehende Babylon* (Berlin, 1913).

Krämer, W. *Antiquity*, xxxiv (1960), 191–200.

Krautheimer, R. *Early Christian and Byzantine Architecture* (London, 1965), pp. 79, 90, 129–30, 180–1, 189, 196, 201–2, 219, 233, 274–80, 294, 343. note 345.

Lambert, E. *Caen roman et gothique* (Paris, 1935).

Lamperez Y. Romea, V. *Historia de la arquitectura cristiana española en ia edad media* (Madrid, 1930).

Lassus, J. B. A., Durand, A. and Durand, P. *Monographie de la cathédrale de Chartres* (Paris, 1881).

Lasteyrie, R. de *L'Architecture religieuse en France à l'epoque romane*. 2nd edition (Paris, 1929), pp. 62, 526.

Leseuer, F. *Les églises de Loir-et-Cher* (Paris, 1969), p. 306.

Lethaby, W. R. and Swainson, H. *The Church of Sancta Sophia* (London, 1894).

Limousin Roman (Paris, 1960), p. 129.

Lloyd, R. and Mellaart, J. *Beycesultan*. Anatolian Studies, v (1955) and vi (1956).

Lopez, R. S. *The Birth of Europe* (London, 1966).

Luca, de (editor). *Architettura medievale armena* (Rome, 1968).

Luc-Benoist (no initial). *Notre-Dame de L'Epine* (Petites Monographies des Grands Edifices de la France, Paris, 1933), p. 37.

Mâle, E. *La fin du paganisme en Gaule et les plus anciennes basiliques chrétiennes* (Paris, 1950).

Maqrizi, Ziada's edition, p. 526.

Mariaclotilae, M. *Architettura romanica comesca*, i (Milan, 1960), pp. 70–75, pl. 82.

Meldahl, F. *Venedig: dets Historie og dets Mindesmaerker* (Copenhagen, 1903), pp. 68–71, fig. 14, pls. 34 and 70.

Mezzanotte, P. and Bascapé, G. C. *Milano nell'arte e nelea storia* (Milan, 1948), p. 7.

Millet, G. *Monuments byzantins de Mistra* (Paris, 1910), pls. 18:1, 32:1, 34:1, 36:1, 38:1, 39:1, 39:3.

Millet, G. *L'Ecole grecque dans l'architecture byzantine* (Paris, 1916).

Millet, G. *L'Ancient art serbe. Les églises* (Paris, 1919), pp. 19, 23, 24, 25, 62, 65, 78, 82.

Mohrmann, K. *Lehrbuch der gotische Konstruktionen von G. Ungewitter* (Leipzig, 1901).

Molajoli, B. *La Basilica Eufrasiana di Parenzo* (Padua, 1943).

Montel, I.-A. *Building Structures in Earthquake Countries* (London, 1912).

Moreau-Nélaton, E. *Les églises des chez nous*. 2 vols. (Soissons, 1913 and 1914), p. 187.

Morgan, W. Ll. *Archaeologia Cambrensis* (1907), 55–56, 138–174.

Muller-Weiner, W. *Castles of the Crusaders* (1966), pp. 78–80, note 72, pl. 48.

Muñoz, A. *Il restauro della basilica di Santa Sabina* (Rome, 1938).

Mylonas, G. E. *Ancient Mycenae* (Princeton, 1957).

Nash-Williams, V. E. *Archaeologia Cambrensis* (1930), 102.

Olivero, E. *L'Antica chiesa di San Constanze sul Monte (Cuneo)* (Turin, 1929).

Olivero, E. *Architettura religiosa preromanica e romanica nell'Archidiocesi di Torino* (Turin, 1940), pp. 173–82.

Oprescu, G. *Bisenche Cetăti Ale Sasitor Din Ardeal* (Editura Academiei Republicii Populare Romine, Bucharest, 1956), pp. 13–15.

Osten, H. H. von der. *Alishar Hüyük*. Oriental Institute Publication No. 29 (Chicago, 1937), p. 292.

Oursel, Ch. *Art roman en Bourgogne* (Dijon and Boston, 1928).

Overseas Building Notes. Earthquake Protection (London, Oct. 1959), pp. 1–22.

Peers, Sir C. 'Recent Discoveries in the Minsters of Ripon and York.' *Antiquaries Journal*, xi (1931), 113–22.

Philo of Byzantium. *Work on architecture*. Published in *Revue de Philologie* (Paris, 1879).

Piggott, S. *Ancient Europe* (Edinburgh, 1965), pp. 204, 210, 216–7.

Plancher, U. *Histoire générale et particulière de Bourgogne*. 4 vols. (Dijon, 1739).

Pococke, R. *Description of the East*. Vol. i (London, 1743).

Porec, C. *La cathédrale d'Auxerre* (Petites Monographies des Grands Cathédrales de la France, Paris, 1926).

Porter, A. K. *Lombardic Architecture*. 3 vols. and atlas (London, 1915–17): i, pp. 94, 288–9, 319; ii, pp. 106, 295, 365, 371; iii, pp. 148, 179, 225, 447, 457, 539.

Porter, A. K. *Medieval Architecture* (New York, 1966), pp. 228, 329.

Previté-Orton, C. W. *The Shorter Cambridge Medieval History*. 2 vols. (Cambridge, 1952).

Procopius. *De Aed*. i. p. 4.

Pryce, F. N. and T. D. *Archaeologia Cambrensis* (1927), 335.

Puhringer, R. *Denkmäler der früh- und hochromanischen Baukunst in Österreich* (Vienna and Leipzig, 1930), p. 5.

Quintavallo, A. G. *San Pietro Barabari in Modena* (Modena, 1965).

Rackman, R. B. 'The Nave of Westminster Abbey'. *Proceedings of the British Academy*, iv (1909–10), 35.

Ramsey, Sir W. and Bell, G. L. *The 1001 Churches* (London, 1909), fig. 327.

de Ranquet, H. and E. 'Origine français du berceau roman.' *Bulletin monumental*, xc (1931).

Renn, D. F. *Norfolk Archaeology*, xxxii (1961), 232–5.

Rey, R. *La cathédrale de Cahors et les origines de l'architecture à coupoles d'Aquitaine* (Paris, 1965).

Ricci, C. *Romanesque Architecture in Italy* (London, 1925), v, pls. 143, 172–4, 176.

Rice, Talbot, D. *Byzantine Art* (Oxford, 1935).

Rice, Talbot, D. *English Art 871–1100* (Oxford, 1952).

Rice, Talbot, D. *The Byzantines* (London, 1962).

Rice, Talbot, D. *Constantinople* (London, 1965), p. 95.

Rivoira, G. T. *Lombardic Architecture*. 2 vols. (Oxford, 1933), i, pp. 39, 126, 152.

Robertson, D. S. *Greek and Roman Architecture* (Cambridge, 1943).

Rosenau, H. *Design and Medieval Architecture* (London, 1934).

Rowe, J. B. *Transactions of the Plymouth Institute*, i (1876–8), 259–74.

Royal Commission on Historical Monuments. *Herefordshire*, pp. 39, 74.

Saint-Paul, A. 'La Cathédrale de Reims au XIIIe siecle'. *Bulletin monumental* (1906), 318.

Salet, F. 'La Madeleine de Vézelay et ses dates de construction'. *Bulletin monumental* (1936), 184.

Sallez, A. 'Le réparation d'un chapiteau de la nef de la cathédrale d'Amiens'. *Les Monuments historiques de la France* (Paris, 1968), pp. 67ff.

Salmi, M. *L'Architettura romanica in Toscania* (Rome, 1927), pp. 14, 15, 18, 20, 21, 30, 32, 43, 47, 48, 57.

Salvadori, A. *101 Buildings to be seen in Venice* (Venice, 1969), pp. 21, 24, 26, 34.

Salzman, L. F. *Sussex Archaeological Collections*, li (1908), 98–114.

Schaffen, E. *Die Kunst der Langobarden in Italien* (Leipzig, 1941), pp. 57–8.

Scheffini, F. *Architettura religiosa preromanica e romanica nell'Archidiocesi de Torino* (Turin, 1946).

Schlink, W. *Zwischen Cluny und Clairvaux* (Berlin, 1970).

Schultz, R. W. and Burnley, S. H. *The Monastery of St Luke of Stiris in Phokis, Greece* (London, 1901).

Seager, R. B. *Gournia, Vasiliki and the other Prehistoric Sites on the Isthmus of Hierapetra, Crete* (Philadelphia, 1908).

Speculum, xxxiv (1959).

Stich, F. *Der gotische Kirchenbau in der Pfalz* (Speyer, 1960), p. 121.

Stikas, E. *L'Eglise byzantine de Christianou* (Paris, 1951).

Stoikov, G. *Boyana Church* (Sofia, 1954), p. 6.

Stokes, A. *Venice. An Aspect of Art* (London, 1945).

Stričević, Dj. *Rapports. XII Congrès International des Etudes Byzantines*, vii. *Ochrid* (1961), p. 176.

Strzygowski, J. *Origin of Christian Church Art* (Oxford, 1923).

Strzygowski, J. *L'ancien art chrétien de Syrie, d'après les découvertes de Vogué et l'expédition de Princeton* (Paris, 1936), p. 171.

Swiechowski, Z. *Architektura na Stasku do Poxony XIII wieku* (no publication source, no date).

Swift, E. H. *Roman Sources of Christian Art* (New York, 1951).

Texier, Ch. *Description of Asia Minor* (Paris, 1862).

Thompson, H. C. 'A Row of Cedar Beams'. *Palestine Exploration Quarterly* (1960–1).

Timmers, J. J. M. *A Handbook of Romanesque Art* (London, 1969).

Touring Club Italiano. *L'Arte nel Medioevo* (Milan, 1964).

Traquair, R. 'The Churches of Western Mani'. Ann. of British School of Athens, xv (1908–9), 177–213.

Truchis, P. de *L'Eglise romane de Bussy-le-Grand* (Congres Archéologique, Avallon, 1907).

Utudjlan, E. *Armenian Architecture, 4th–17th centuries*. Translated by Geoffrey Capner (Paris, 1968).

Vallery-Radot, J. *Saint-Philibert de Tournus* (Paris, 1955).

Van der Mear, S. *Cathédrales méconnues de France* (Brussels-Paris, 1968), pp. 81–8.

Van Millingen, A. *Byzantine Churches in Constantinople* (London, 1912), pp. 9, 10, 332.

Vassas, R. 'Travaux à la Madeleine de Vézelay'. *Les Monuments Historiques de la France* (Paris, 1968).

Venditti, A. *Architettura bizantina nell'Italia meridionale*. 2 vols. (Naples, 1968), ii, p. 180.

Victoria County History of Essex, iii (1963), p. 53 and fig. 14.

Viollet-le-Duc, M. *Rational Building*. Translation by G. M. Huss (New York, 1895), pp. 396–400.

Virey, J. 'Les dates de construction de Saint-Philibert de Tournus'. *Bulletin monumental* (1903), 74–9, 532ff.

Volbach, W. F. *Early Christian Art* (London, 1961).

Von Veltheim, H. H. *Burgundische Kleinkirchen bis 1200* (Berlin, 1913).

Wace, A. J. B. *Mycenae* (Princeton, 1949).

Waerm, C. *Medieval Sicily* (London, 1910), pp. 171–8.

Ward-Perkins, J. B. 'The Italian element in late Roman and early medieval architecture'. *Proceedings of the British Academy*, xxxiii (1947), 163–94.

Wheeler, R. E. M. *Archaeological Journal*, xxxvi (1929), 52 and fig. 1.

Wheeler, R. E. M. 'Notes on building construction in Roman Britain'. *Journal of Roman Studies*, xxii (1932), 116–34.

Wheeler, R. E. M. *Maiden Castle* (Society of Antiquaries Research Report xii, 1943).

White, H. J. E. 'The Monasteries of the Wady-el-Natrun'. *Bulletin of the Museum of New York*, xv (1920) and xvi (1921).

White, J. *Art and Architecture in Italy* (London, 1966), pp. 8, 18–19, 20, 184, 191, 192, 345.

Whitehill, W. M. *Spanish Romanesque Architecture of the 11th century* (Oxford and London, 1941).

Whittingham, A. B. *Archaeological Journal*, cvi (1949).

Wilcox, R. 'Excavations at Castle Acre Priory, Norfolk'. *Norfolk Archaeology*, 1981.

Willemsens, C. A. and Odenthal, D. *Apulia* (London, 1959), pp. 22–4, 31, 37–8, 40.

Willson, A. *Proceedings of the Archaeological Institute* (1848), 285–6.

Wilson, A. E. *Sussex Archaeological Collections*, lxxxix (1938) and xc (1939).

Woolley, Sir C. L. *Carchemish*. Part 2 (London, 1921).

Woolley, Sir C. L. *Excavations at Ur* (London, 1954).

Wooley, Sir C. L. *Alalakh* (London, 1955).